Disclaimer

The publisher of this book is by no way associated with the National Institute of Standards and Technology (NIST). The NIST did not publish this book. It was published by 50 page publications under the public domain license.

50 Page Publications.

Book Title: Effect of Bulk Lubricant Concentration on the Excess Surface Density During R134a Pool Boiling with Extensive Measurement and Analysis Details

Book Author: Mark A. Kedzierski

Book Abstract: This paper investigates the effect that the bulk lubricant concentration has on the non-adiabatic lubricant excess surface density on a roughened, horizontal flat pool-boiling surface. Both pool boiling heat transfer data and lubricant excess surface density data are given for pure R134a and three different mixtures of R134a and a polyolester lubricant (POE). A spectrofluorometer was used to measure the lubricant excess density that was established by the boiling of a R134a/POE lubricant mixture on a test surface. The lubricant is preferentially drawn out of the bulk refrigerant/lubricant mixture by the boiling process and accumulates on the surface in excess of the bulk concentration. The excess lubricant resides in an approximately 40 mm layer on the surface and influences the boiling performance. The lubricant excess surface density measurements were used to modify an existing dimensionless excess surface density parameter so that it is valid for different reduced pressures. The dimensionless parameter is a key component for a refrigerant/lubricant pool boiling model given in the literature. In support of improving the boiling model, both the excess measurements and heat transfer data are provided for pure R134a and three R134a/lubricant mixtures at 277.6 K. The heat transfer data shows that the lubricant excess layer causes an average enhancement of the heat flux of approximately 50 % for the 0.5 % lubricant mass fraction mixture relative to pure R134a heat fluxes between 4 kW/m2 and 20 kW/m2. Conversely, both the 1 % and the 2 % lubricant mass fraction mixtures experienced an average degradation of approximately 60 % in the heat flux relative to pure R134a heat fluxes between approximately 4 kW/m2 and 20 kW/m2. This study is an effort toward generating data that can be used to support a boiling model that can be used to predict whether lubricants degrade or improve boiling performance.

Citation: NIST Interagency/Internal Report (NISTIR)

Keyword: adsorption;alternative refrigerants;boiling;enhanced heat;fluorescence;non-adiabatic lubricant excess surface;pool boiling;R134a;refrigerant/lubricant mixtures

NISTIR 7051

Effect of Bulk Lubricant Concentration on the Excess Surface Density During R134a Pool Boiling with Extensive Measurement and Analysis Details

Mark A. Kedzierski

U.S DEPARTMENT OF COMMERCE
National Institute of Standard and Technology
Building Environment Division
Building and Fire Research Laboratory
Gaithersburg, MD 20899-8631

September 2003

U.S. Department of Commerce
Donald L. Evans, Secretary

National Institute of Standards and Technology
Arden l. Bement, Jr., Director

Effect of Bulk Lubricant Concentration on the Excess Surface Density During R134a Pool Boiling with Extensive Measurement and Analysis Details

M. A. Kedzierski
National Institute of Standards and Technology
Bldg. 226, Rm B114
Gaithersburg, MD 20899
Phone: (301) 975-5282
Fax: (301) 975-8973

ABSTRACT

This paper investigates the effect that the bulk lubricant concentration has on the non-adiabatic lubricant excess surface density on a roughened, horizontal flat pool-boiling surface. Both pool boiling heat transfer data and lubricant excess surface density data are given for pure R134a and three different mixtures of R134a and a polyolester lubricant (POE). A spectrofluorometer was used to measure the lubricant excess density that was established by the boiling of a R134a/POE lubricant mixture on a test surface. The lubricant is preferentially drawn out of the bulk refrigerant/lubricant mixture by the boiling process and accumulates on the surface in excess of the bulk concentration. The excess lubricant resides in an approximately 40 µm layer on the surface and influences the boiling performance. The lubricant excess surface density measurements were used to modify an existing dimensionless excess surface density parameter so that it is valid for different reduced pressures. The dimensionless parameter is a key component for a refrigerant/lubricant pool-boiling model given in the literature. In support of improving the boiling model, both the excess measurements and heat transfer data are provided for pure R134a and three R134a/lubricant mixtures at 277.6 K. The heat transfer data shows that the lubricant excess layer causes an average enhancement of the heat flux of approximately 24 % for the 0.5 % lubricant mass fraction mixture relative to pure R134a heat fluxes between 5 kW/m^2 and 20 kW/m^2. Conversely, both the 1 % and the 2 % lubricant mass fraction mixtures experienced an average degradation of approximately 60 % in the heat flux relative to pure R134a heat fluxes between approximately 4 kW/m^2 and 20 kW/m^2. This study is an effort toward generating data that can be used to support a boiling model that can be used to predict whether lubricants degrade or improve boiling performance.

Keywords: adsorption, alternative refrigerants, boiling, enhanced heat transfer, fluorescence, non-adiabatic lubricant excess surface density, pool boiling, R134a, refrigerant/lubricant mixtures, smooth surface, surfactant[1]

---

[1] Certain trade names and company products are mentioned in the text or identified in an illustration in order to adequately specify the experimental procedure and equipment used. In no case does such an identification imply recommendation or endorsement by the National Institute of Standards and Technology, nor does it imply that the products are necessarily the best available for the purpose.

INTRODUCTION

The addition of lubricant to refrigerant can significantly alter the boiling performance due to lubricant accumulation at the heat transfer surface. Stephan (1963) was one of the first researchers to note that a lubricant-rich layer exists near the tube wall. The excess concentration (excess surface density) arises from the low vapor pressure of the lubricant relative to the refrigerant. The lubricant can be locally drawn out of solution as a consequence of refrigerant evaporation at the heat transfer surface. The refrigerant/lubricant liquid mixture travels to the heated wall, and the refrigerant preferentially evaporates from the surface leaving behind a liquid phase enriched in lubricant. A balance between deposition and removal of the lubricant establishes the thickness of the excess lubricant at the surface. It is hypothesized that the lubricant excess layer controls the bubble size, the site density and, in turn, the magnitude of the heat transfer.

Kedzierski (2002a) developed a fluorescence measurement technique to verify the existence of the lubricant excess layer during pool boiling. A spectrofluorometer was specially adapted for use with a bifurcated optical bundle so that fluorescence measurements could be made perpendicular to the heat transfer surface. The study suggested that the excess layer was pure lubricant with a thickness ranging from 0.04 mm to 0.06 mm depending on the heat flux. The study examined only one R123/mineral oil mixture. Kedzierski (2002b) expanded the study by using the new technique to investigate the effect of three R123/mineral oil bulk concentrations on a mineral oil excess layer. The data for the three mixtures led to the development of a semi-theoretical model for predicting R123/lubricant mixture pool boiling heat transfer Kedzierski (2003). The model relies on excess surface density measurements and a dimensionless parameter representing the excess measurements. The present study uses the fluorescence measurement technique to extend the database to three R134a/polyolester lubricant (POE) mixtures to test and extend the dimensionless lubricant excess surface density parameter to other refrigerants and lubricants. Three different POE (DE589[2]) mass compositions were investigated: 99.5/0.50, 99.02/0.98, and 98.04/1.96 (nominally 99.5/0.5, 99/1, and 98/2). The DE589 POE lubricant has a viscosity of 22 $\mu m^2/s$ at 313.15 K. The lubricant was chosen for its somewhat favorable fluorescence characteristics and its commercial use with R134a.

APPARATUS

Figure 1 shows a schematic of the apparatus that was used to measure the pool boiling data of this study. More specifically, the apparatus was used to measure the liquid saturation temperature (T_s), the average pool-boiling heat flux (q''), the wall temperature (T_w) of the test surface, and the fluorescence intensity from the boiling surface (F). The three principal components of the apparatus were test chamber, condenser, and purger. The internal dimensions of the test chamber were 25.4 mm × 257 mm × 1.54 m. The test chamber was charged with approximately 7 kg of R134a from the purger, giving a liquid height of approximately 80 mm above the test surface. As shown in Fig. 1, the test section was visible through two opposing, flat 150 mm × 200 mm quartz windows. The bottom of the test surface was heated with high velocity (2.5 m/s) water flow. The vapor

---

[2] ICI's EMKARATE RL DE 589

produced by liquid boiling on the test surface was condensed by the brine-cooled, shell-and-tube condenser and returned as liquid to the pool by gravity.

Figure 1 also shows the spectrofluorometer that was used to make the fluorescence measurements and the fluorescence probe perpendicular to the heat transfer surface. The fluorescence probe was a bifurcated optical bundle with 168 fibers spanning from the spectrofluorometer to the test surface. The 168 fibers of the probe were split evenly between the fibers to transmit the incident intensity (I_o) to the test surface and those to receive the fluorescence intensity (F) from the lubricant on the test surface. Further details of the test apparatus can be found in Kedzierski (2002a) and Kedzierski (2001a).

TEST SURFACE
Figure 2 shows the oxygen-free high-conductivity (OFHC) copper flat test plate used in this study. The test plate was machined out of a single piece of OFHC copper by electric discharge machining (EDM). OFHC copper was chosen because of its well-known thermal conductivity and because its oxidation and wetting characteristics are expected to be similar to copper alloys used commercially with refrigerants. A tub grinder was used to finish the heat transfer surface of the test plate with a crosshatch pattern. Average roughness measurements were used to estimate the range of average cavity radii for the surface to be between 12 µm and 35 µm. The relative standard uncertainty of the cavity measurements were approximately ± 12 %. Further information on the surface characterization can be found in Kedzierski (2001a).

MEASUREMENTS AND UNCERTAINTIES
The standard uncertainty (u_i) is the positive square root of the estimated variance u_i^2. The individual standard uncertainties are combined to obtain the expanded uncertainty (U), which is calculated from the law of propagation of uncertainty with a coverage factor. All measurement uncertainties are reported for a 95 % confidence interval except where specified otherwise. For the sake of brevity, only an outline of the basic measurements and uncertainties are given below. Complete detail on the heat transfer measurement techniques and uncertainties can be found in Kedzierski (2000 and 2001b).

Heat Transfer
All of the copper-constantan thermocouples and the data acquisition system were calibrated against a glass-rod standard platinum resistance thermometer (SPRT) and a reference voltage to a residual standard deviation of 0.005 K. Considering the fluctuations in the saturation temperature during the test and the standard uncertainties in the calibration, the expanded uncertainty of the average saturation temperature was no greater than 0.04 K. Consequently, it is believed that the expanded uncertainty of the temperature measurements was less than 0.1 K.

Twenty 0.5 mm diameter thermocouples were force fitted into the wells of the side of the test plate shown in Fig. 2. The heat flux and the wall temperature were obtained by regressing the measured temperature distribution of the block to the governing two-dimensional conduction equation (Laplace equation). In other words, rather than using the boundary conditions to solve for the interior temperatures, the interior temperatures were used to solve for the boundary conditions following a backward stepwise procedure given in Kedzierski (1995). Fourier's law and the fitted constants from the Laplace equation were used to calculate the average heat flux (q'') normal to and evaluated at the heat transfer surface based on its projected area. The average wall temperature (T_w) was calculated by integrating the local wall temperature (T). The wall superheat was calculated from T_w and the measured temperature of the saturated liquid (T_s). Considering this, the relative expanded uncertainty in the heat flux ($U_{q''}$) was greatest at the lowest heat fluxes, approaching 8 % of the measurement at 10 kW/m^2. In general, the $U_{q''}$ was relatively constant between 4 % and 5 % for heat fluxes above 25 kW/m^2. The average random error in the wall superheat (U_{Tw}) was between 0.02 K and 0.08 K. Plots of $U_{q''}$ and U_{Tw} versus heat flux can be found in Appendix A.

Fluorescence
Kedzierski (2002a) describes the method for calibrating the emission intensity measured with the spectrofluorometer and the bifurcated optical bundle as shown in Fig. 1 against the bulk lubricant mass fraction. As outlined in Appendix B, the excitation and emission wavelengths for the spectrofluorometer with experimentally determined as 394 nm and 467 nm, respectively. One modification was done to the fluorescence measurement method of the previous study. Because the fluorescence intensity of the present POE was significantly less than that of the previous study with mineral oil, signal noise from stray wavelengths was of the order of the POE emission. To remedy this, the spectrofluorometer was modified such that both the excitation and the emission were limited to a narrow wavelength band with interference filters (see Appendix C).

Figure 3 shows eight different calibration runs using the calibration procedure described above. The solid line depicts the regression of the intensity of the fluorescence emission (F) to the Beer-Lambert-Bougher law (Amadeo et al., 1971) as a function of the bulk lubricant mass fraction (x_b) and the bulk liquid mixture density (ρ_b):

$$F_c = 518\left[1 - 10^{-9.43 \times 10^{-5} x_b \rho_b}\right] \quad (1)$$

The average 95 % confidence interval for the lubricant mass fraction is approximately ±0.01. The width of the confidence interval is a function of the lubricant fluorescence. A greater absolute fluorescence intensity would reduce the scatter in the data. The fluorescence calibration measurements that were used to generate Eq. (1) are given in Appendix D. The pure lubricant liquid density was measured in a pycnometer and is given in Appendix E. The mixture densities were calculated on a linear mass weighted basis.

Because the molar mass of the lubricant is unknown, the surface excess density (Γ) is defined in this work on a mass basis as:

$$\Gamma = \rho_e x_e l_e - \rho_b x_b l_e \quad (2)$$

where the l_e is the thickness of the lubricant excess layer. Precedence for reporting the surface excess density in mass units is given by citing the work of McBain and

Humphreys (1932) in which they experimentally verified the Gibbs adsorption equation. A non-zero value of Γ implies that an excess layer exists on the surface.

The equation for calculating the surface excess density from the measured fluorescence emission intensity (F_m) (Kedzierski, 2002b) for the DE589 lubricant was slightly modified (see Appendix F) to account for the temperature difference between the excess layer and the bulk fluid:

$$\Gamma = \rho_e x_e l_e - \rho_b x_b l_e = \frac{\rho_b x_b \left(\frac{\rho_{L,T_e}}{\rho_{L,T_b}} - \frac{\rho_b x_b}{\rho_{L,T_b}} \right)\left(\frac{F_m}{F_c} - 1 \right)}{\frac{I_{oe}}{I_{ob}}\left(1 + 1.165 \frac{\varepsilon}{M_L} x_b \rho_b l_b \right)\frac{e^{\beta(T_e - T_b)}}{l_b} - 1.165 \frac{\varepsilon}{M_L} x_b \rho_b \left(\frac{F_m}{F_c} - 1 \right)} \quad (3)$$

where the value of $\frac{\varepsilon}{M_L}$ was obtained from the fluorescence calibration as 0.0646 m²/kg, and the fluorescence temperature dependence coefficient (β) was experimentally determined to be 0.01 K⁻¹ (Appendix G). All of the fluid properties are evaluated at the bulk fluid temperature (T_b) with the exception of the $\rho_{L,Te}$, which is the pure lubricant density evaluated at the average temperature of the excess layer (T_e). If T_e and T_b are equal, Eq. (6) reduces to the original form that was given in Kedzierski (2002b). For the measurements taken for this study, the correction to account for the temperature dependence of the fluorescence in the excess layer effected Γ by as much as 2 %.

Input for Eq. (3) is as follows. The fluorescent intensity from the calibration (F_c) is obtained from Eq. (1) evaluated at the charged bulk lubricant concentration of test fluid in the boiling apparatus. The l_b is the distance between the probe and the heat transfer surface and $l_b >> l_e$. The density of the pure lubricant is ρ_L. The ratio of the absorption of the incident excitation in the bulk to that in the excess layer (I_{oe}/I_{ob}) was obtained from the measured absorption spectrum of a 92.9/7.1 mass fraction mixture of R134a and DE589 shown in Fig. H.1 of Appendix H. Absorption ratios for the 99.5/0.5, the 99/1.0, and the 98/2.0 mixtures were 0.95, 0.993, 0.986, respectively. A sample calculation of the absorption ratio for the 0.02 lubricant mass fraction mixture is given in Appendix H.

Equation (3) was derived while assuming that the excess layer exists at a minimum thickness, i.e., the excess layer is entirely lubricant. Small excess layer mass fractions give excess layers that are unrealistically too thick. For example, the excess layer thickness ranges from 0.7 mm to 1.3 mm for an assumed excess layer mass fraction of 0.03. Two physical mechanisms support a thin, pure lubricant layer: (1) the preferential evaporation of the refrigerant tends to enrich the excess layer in the lubricant phase; while (2) the bubble pumping action of lubricant from the surface tends to minimize the thickness of the lubricant excess layer.

EXPERIMENTAL RESULTS

Heat Transfer

The heat flux was varied between the range of 130 kW/m^2 and 5 kW/m^2 to simulate most possible operating conditions for R134a chillers. All pool-boiling tests were taken at 277.6 K saturated conditions. The data were recorded consecutively starting at the largest heat flux and descending in intervals of approximately 4 kW/m^2. The descending heat flux procedure minimized the possibility of any hysteresis effects on the data, which would have made the data sensitive to the initial operating conditions. Table 2 presents the measured heat flux and wall superheat for all the data of this study. Table 3 gives the number of test days and data points for each fluid.

The R134a/mixture was prepared by charging the test chamber (see Fig. 1) with pure R134a to a known mass. Next, a measured mass of DE589 was injected with a syringe through a port in the test chamber. The lubricant was mixed with R134a by flushing pure R134a through the same port where the lubricant was injected. All compositions were determined from the masses of the charged components and are given on a mass percent basis. The maximum uncertainty of the composition measurement is approximately 0.02 %, e.g. the range of a 2.0 % composition is between 1.98 % and 2.02 %.

Figure 4 is a plot of the measured heat flux (q'') versus the measured wall superheat ($T_w - T_s$) for pure R134a at a saturation temperature of 277.6 K. The closed circles represent six days of boiling measurements made over a period of approximately two weeks. "Break-in" data taken during the first three days of testing, before the surface had fully "aged," are not shown on the figure, but are given in Appendix I. The aging effect has previously been observed for this surface for the tests immediately following cleaning and installation (Kedzierski, 2001b). The present surface was cleaned prior to installation in the test apparatus sequentially with acetone, Tarnex^{TM}, hot tap water, and acetone. Marto and Lepere (1982) have also observed a surface aging effect on pool boiling data that was sensitive to initial surface conditioning and fluid properties.

The solid lines shown in Fig. 4 are cubic best-fit regressions or estimated means of the data. Nine of the 144 measurements were removed before fitting because they were identified as "outliers" based on having both high influence and high leverage (Belsley et al., 1980). Table 4 gives the constants for the cubic regression of the superheat versus the heat flux for all of the fluids tested here. The residual standard deviation of the regressions - representing the proximity of the data to the mean - are given in Table 5. The dashed lines to either side of the mean represent the lower and upper 95 % simultaneous (multiple-use) confidence intervals for the mean. From the confidence intervals, the expanded uncertainty of the estimated mean wall superheat was 0.1 K and 0.04 K for superheats less than and greater than 6 K, respectively. Table 6 provides the average mean wall uncertainty for all of the test data.

Figure 4 also provides the smooth tube boiling data of Webb and Pais (1992) at the same saturation temperature as the present tests. The largest differences between the Webb and Pais (1992) smooth tube and the present flat plate measurements are found at the extremes of the data set. For the same superheat, the Webb and Pais (1992) smooth tube heat flux is 60 % greater than and 40 % less than the measured heat flux for the flat plate at 10 kW/m^2 and 70 kW/m^2, respectively. Averaged over the entire heat flux range of the data, the Webb and Pais (1992) heat flux is 18 % less than the present flat plate heat flux for the same superheat. For further reference, Fig. 4 shows the predictions from a free convection correlation for a horizontal plate with the heated surface facing upward, which was recommended by Incropera and Dewitt (1985). The natural convection heat flux predictions range between 10 % to 4 % of the boiling heat fluxes for the same wall superheat.

Figure 5 plots the measured heat flux (q'') versus the measured wall superheat ($T_w - T_s$) at a saturation temperature of 277.6 K for the (99.5/0.5), (99/1), and the (98/2) R134a/DE589 mixtures. The mean of the pure R134a "aged data" is plotted as a dashed line. Comparison of the 99.5/0.5 mixture boiling curve to the mean R134a boiling curve shows that they intersect at a superheat of approximately 5.8 K. For mean superheats between 4 K and 5.8 K, the 99.5/0.5 mixture exhibits an enhancement in the heat flux as compared to the pure refrigerant. In contrast, the pure R134a heat flux is greater than that of the 99.5/0.5 mixture for superheats greater than 5.8 K. Apparently, the lubricant enhances the site density and, in turn, the heat transfer for superheats between 4 K and 5.8 K. This enhancement mechanism is ineffective at superheats greater than 5.8 K because nearly all of the available sites have been activated leaving no opportunity for improvement. For superheats greater than 5.8 K, the degradation exhibit by the 99.5/0.5 mixture results from the decreased in bubble size associated with the lubricant.

Figure 5 shows that the boiling performance of the 99/1 mixture for all superheats is less than that of both pure R134a and the 99.5/0.5 mixture. In addition, the scatter in the 99/1 mixture data is marginally larger than that for the 99.5/0.5 mixture given that the residual standard deviation is 0.19 K and 0.26 K, respectively. The scatter of the 98/2 mixture is the largest of the fluids tested being 0.39 K. The general trend of increasing measurement variability with increasing lubricant concentration is consistent with that observed by Kedzierski (2002b and 2001b). The boiling performance of the 98/2 mixture does not follow the trend of decreasing performance with increasing lubricant mass fraction for heat fluxes greater than approximately 30 kW/m^2. For a given superheat for heat fluxes greater than 30 kW/m^2, the heat flux of the 98/2 mixture is between that of the 99.5/0.5 and the 99/1 mixtures. This characteristic is not likely due to the larger heat flux range of the 98/2 measurements because some of the data with the most favorable performance was taken at a starting heat flux of 80 kW/m^2, which coincides with the starting heat flux of much of the 99/1 mixture data. The observed enhancement associated with the increase in lubricant mass fraction to 2 % may likely have been induced by a significant increase in site density without a significant loss in bubble size as compared to the 1 % lubricant concentration.

A more detailed comparison of the mixture and the pure fluid heat transfer performance is given in Fig. 6. Figure 6 plots the ratio of the mixture to the pure R134a heat flux (q''_m/q''_p) versus the pure R134a heat flux (q''_p) at the same wall superheat. A heat transfer enhancement exists where the heat flux ratio is greater than one and the 95 %

simultaneous confidence intervals (depicted by the shaded regions) do not include the value one. Figure 6 shows that the R134a/ DE589 (99.5/0.5) mixture exhibits an enhancement over pure R134a for heat fluxes between approximately 7 kW/m^2 and 22 kW/m^2. The maximum heat flux ratio for the 99.5/0.5 mixture was 1.36 ± 0.04 at 12.5 kW/m^2. The average heat flux ratio for the R134a/ DE589 (99.5/0.5) mixture from approximately 6 kW/m^2 to 21 kW/m^2 was 1.24. The average heat flux ratio from approximately 6 kW/m^2 to 81 kW/m^2 was 0.84.

Figure 6 shows that the R134a/ DE589 (99/1) mixture exhibits a heat transfer degradation for all heat fluxes shown. The maximum heat flux ratio for the 99/1 mixture was 0.65 ± 0.24 at a pure R134a heat flux of 15 kW/m^2. The average heat flux ratio for the R134a/ DE589 (99/1) mixture from approximately 15 kW/m^2 to 82 kW/m^2 was 0.4.

Figure 6 shows that the R134a/ DE589 (98/2) mixture exhibits a degradation for all the heat fluxes that were tested. The maximum heat flux ratio of 0.82 ± 0.04 was observed at pure R134a heat flux of 6.1 kW/m^2. The average heat flux ratio for the R134a/DE589 (98/2) mixture from approximately 6 kW/m^2 to 88 kW/m^2 was 0.34.

For the comparisons made in Fig. 6, the available range of pure R134a heat fluxes corresponds to heat fluxes less than 50 kW/m^2 for the 99.5/0.5 mixture and to heat fluxes less than 25 kW/m^2 for the 99/1 and the 98/2 mixtures. In order to examine the relative magnitudes of the mixture heat fluxes greater than 50 kW/m^2, Fig. 7 normalizes the mixture heat flux relative to that of the 98/2 mixture ($q''_{2\%}$) rather than pure R134a heat flux (at the same wall superheat). Figure 7 illustrates that the boiling performance of the 99.5/0.5 mixture averaged between 5 kW/m^2 and 60 kW/m^2 is approximately 94 % greater than that of the 98/2 mixture. The maximum heat flux ratio for the 99.5/0.5 mixture relative to the 98/2 mixture is 2.6 ± 0.01 at 10 kW/m^2. The heat flux ratio of the 99/1 mixture has a maximum of 1.35 ± 0.09 at 11 kW/m^2 and an average of approximately 0.91 between 8 kW/m^2 and 65 kW/m^2.

Fluorescence
Although the heat flux was varied between the range of 130 kW/m^2 and 5 kW/m^2, fluorescence measurements were limited between 50 kW/m^2 and 15 kW/m^2 to limit the time required to quench the boiling below the fluorescence probe. Bubbles under the probe would have misdirected the excitation and the emission lights resulting in a significant reduction in the emission signal. Consequently, the boiling was quenched prior to making fluorescent measurements. Fluorescent measurements were made with respect to time from that when no bubbles were visible below the probe. These measurements were extrapolated to just before quenching to obtain the fluorescence of the surface during boiling just prior to quenching. The difference between the extrapolated fluorescence and the fluorescence averaged over the measurement time period was, in general, less than the scatter between measurements.

Figure 8 shows the lubricant excess surface density measurements for the three R134a/DE589 mixtures as calculated with Eq. (3) versus the an excess property group developed in Kedzierski (2003) from R123/York-C Γ measurements taken in Kedzierski (2002b). Table 7 provides the raw fluorescence intensity measurements that were used in Eq. (3). The excess property group was used to derive the following constant for the dimensionless Γ:

$$\frac{(\rho_L - \rho_b x_b) x_b T_s \sigma}{(1-x_b)\rho_L h_{fg} \Delta T_s \Gamma} = 5.9 \times 10^{-7} \qquad (4)$$

Equation (4) was a key component of the refrigerant/lubricant pool boiling heat transfer model present in Kedzierski (2003) and essentially represents the slope of the R123/York-C data in Fig. 8. Although, Eq. (4) agrees well with the R123/York-C measurements, it fails to predict the Γ measurements for the R134a/DE589 (99/1) and (98/2) mixtures. A probable reason for this is that larger reduced pressure, and thus, smaller diameter bubbles associated with R134a as compared to R123 cannot remove the lubricant from the wall as well as the larger R123 bubbles. The consequence of the reduced effectiveness of lubricant removal is that the R134a Γ is larger than what would be expected for R123 for all other conditions given in Eq. (4) being fixed. Consequently, it is likely that Eq. (4) should include a reduced pressure term to account for the reduced lubricant removal effectiveness at larger reduced pressure.

In light of this, two modifications were made to the dimensionless Γ given in Eq. (4) so that both R123 and R134a Γ measurements could be predicted with a single relationship. First a reduced pressure ($P_r = P/P_c$) term similar to that given by Semeria (1962) and Nishikawa and Urakwa (1960) was included to capture the effects of pressure on boiling. The second modification was to allow the thickness of the lubricant on the bubble to vary with bulk lubricant mass fraction. Because the dimensionless Γ was developed from a physical model based on a lubricant mass balance on the bubble, the thickness of the lubricant on the bubble is of primary importance. In the original model, the thickness was assumed to be constant. The present modification allows for a small variation of the lubricant thickness on the bubble with respect to mass fraction. The complete derivation including the present modifications to the dimensionless Γ is given in Appendix J. The modified dimensionless Γ which accounts for the effects of reduced pressure and variable lubricant thickness on the bubble is:

$$\frac{(\rho_L - \rho_b x_b) x_b^{1.8} T_s \sigma P_r}{(1-x_b)\rho_L h_{fg} \Delta T_s \Gamma} = 2.8 \times 10^{-10} \pm 0.2 \times 10^{-10} \qquad (5)$$

Equation (5) was obtained from a single regression of the present R134a/DE589 and the R123/York-C measurements from Kedzierski (2002b). Figure 9 shows that Eq. (5) represents the measured Γ for both the R134a/DE589 and the R123/York-C mixtures to an average residual standard deviation of 0.1 kg/m^2. Consequently, the modified dimensionless Γ can be used to broaden the applicability of the pool boiling model to refrigerants and lubricants other than R123 and York-C.

CONCLUSIONS

A newly developed fluorescent measurement technique was used to investigate the effect of bulk lubricant concentration on the lubricant excess layer during boiling of R134a and a polyolester lubricant (DE589). A spectrofluorometer was specially adapted for use with a bifurcated optical bundle so that fluorescence measurements could be made perpendicular to the heat transfer surface. The heat transfer surface was a horizontal, roughened, copper flat plate. The lubricant excess surface density was shown to increase with respect to a modified dimensionless lubricant excess surface density parameter.

The boiling heat transfer measurements were simultaneously taken with the fluorescence measurements. The heat transfer data shows that the lubricant excess layer causes an average enhancement of the heat flux of approximately 24 % for the 0.5 % lubricant mass fraction mixture relative to pure R134a heat fluxes between 5 kW/m^2 and 20 kW/m^2. Conversely, both the 1 % and the 2 % lubricant mass fraction mixtures experienced an average degradation of approximately 60 % in the heat flux relative to pure R134a heat fluxes between approximately 4 kW/m^2 and 20 kW/m^2.

The lubricant excess surface density measurements were used to modify an existing dimensionless excess surface density parameter so that it is valid for different reduced pressures. The dimensionless parameter is a key component for a refrigerant/lubricant pool-boiling model given in the literature by the author. In support of improving the boiling model, both the excess measurements and heat transfer data are provided for pure R134a and three R134a/lubricant mixtures at 277.6 K. This study is an effort toward generating data that can be used to support a boiling model that can be used to predict whether lubricants degrade or improve boiling performance.

ACKNOWLEDGEMENTS

This work was jointly funded by NIST and the U.S. Department of Energy (project no. DE-01-95CE23808.000 modification #A004) under Project Manager Arun Vohra. Thanks go to the following people for their constructive criticism of the first draft of the manuscript: Mr. J. Bogart, Mr. W. Payne, and Dr. P. Domanski. The author would also like to express appreciation to Mr. J. Fry, and Mr. G. Glaeser of the NIST Building Environment Division for data collection. Furthermore, the author extends appreciation to Dr. W. Guthrie and Mr. A. Heckert of the NIST Statistical Engineering Division for their consultations on the uncertainty analysis. Special thanks goes to Dr. T. Vorburger of the NIST Precision Engineering Division for making the roughness measurements of the crosshatch surface. The DE589 lubricant that was donated by Dr. S. Randles and Dr. T. Dekleva of ICI is much appreciated.

NOMENCLATURE

English Symbols

A	absorbance
A_n	regression constant in Table 4 n=0,1,2,3
c	concentration, mol/m^3
C_1	constant defined in Eq. (G.1)
D	constant defined in Eq. (J.7)
F	fluorescence intensity
F_c	fluorescence intensity from calibration (Eq. 1)
F_m	fluorescence intensity measured from boiling surface
h_{fg}	latent heat of vaporization of refrigerant, kJ/kg
I_o	incident intensity, V
I_t	transmitted intensity, V
K	constant defined in Eq. (J.2)
k	thermal conductivity, W/m·K
l	path length, m
l_a	thickness of lubricant bubble cap, m
l_e	thickness of excess layer, m
L_y	length of test surface, m
M_L	molar mass of lubricant, kg/mol
m	mass, kg
n	constant defined in Eq. (J.2)
P	vapor pressure, kPa
P_c	critical pressure, kPa
P_r	reduced pressure (P/P_c), kPa
p	constant defined in Eq. (J.5)
q''	average wall heat flux, W/m^2
r_b	bubble departure radius, m
Ra_L	Rayleigh number based on projected area (Fig. 4)
T	temperature, K
T_w	temperature at roughened surface, K
U	expanded uncertainty
u_i	standard uncertainty
x	mass fraction of lubricant
X	model terms given in Table 1
y	test surface coordinate in Fig. 4, m
z	test surface coordinate in Fig. 4, m

Greek symbols

β	temperature dependence of fluorescence coefficient, K^{-1}
Γ	lubricant excess surface excess, kg/m^2
γ	constant defined in Eq. (J.5)
ΔT_s	wall superheat: $T_w - T_s$, K
ε	extinction coefficient, m^2/mol
ζ	fraction of excess layer removed per bubble

ρ	mass density of liquid, kg/m^3
σ	surface tension of refrigerant, kg/s^2
Φ	quantum efficiency of fluorescence

English Subscripts

b	bulk
e	excess layer
L	lubricant
m	measured, mixture
p	pure R134a
q"	heat flux
s	saturated state
Tw	wall temperature
v	vapor

Superscripts

-	average

REFERENCES

Amadeo, J. P., Rosén C., and Pasby, T. L., 1971, Fluorescence Spectroscopy An Introduction for Biology and Medicine, Marcel Dekker, Inc., New York, p. 153.

Belsley, D. A., Kuh, E., and Welsch, R. E., 1980, Regression Diagnostics: Identifying Influential Data and Sources of Collinearity, New York: Wiley.

Guilbault, G. G., 1967, Fluorescence: Theory, Instrumentation, and Practice, Edward Arnold LTD., London, pp. 91-95.

Incropera, F. P., and DeWitt, D. P., 1985, Fundamentals of Heat and Mass Transfer, 2nd ed., John Wiley & Sons, New York.

Kedzierski, M. A., 2003, "A Semi-Theoretical Model for Predicting R123/Lubricant Mixture Pool Boiling Heat Transfer," Int. J. Refrigeration, Vol. 26, pp. 337-348.

Kedzierski, M. A., 2002a, " Use of Fluorescence to Measure the Lubricant Excess Surface Density During Pool Boiling," Int. J. Refrigeration, Vol. 25, pp.1110-1122.

Kedzierski, M. A., 2002b, " Effect of Bulk Lubricant Concentration on the Excess Surface Density During R123 Pool Boiling," Int. J. Refrigeration, Vol. 25, pp. 1062-1071.

Kedzierski, M. A., 2001a, " Use of Fluorescence to Measure the Lubricant Excess Surface Density During Pool Boiling," NISTIR 6727, U.S. Department of Commerce, Washington, D.C.

Kedzierski, M. A., 2001b, " Effect of Bulk Lubricant Concentration on the Excess Surface Density During R123 Pool Boiling," NISTIR 6754, U.S. Department of Commerce, Washington, D.C.

Kedzierski, M. A., 2000, "Enhancement of R123 Pool Boiling by the Addition of Hydrocarbons," Int. J. Refrigeration, Vol. 23, pp. 89-100.

Kedzierski, M. A., 1995, "Calorimetric and Visual Measurements of R123 Pool Boiling on Four Enhanced Surfaces," NISTIR 5732, U.S. Department of Commerce, Washington.

Nishikawa, K., and Urakawa, K., 1960, "An Experiment of Nucleate Boiling Under Reduced Pressure," *Memoirs of Faculty of Eng.*, Kyushu Univ., Vol. 19, No. 3, pp. 63-71.

Marto, P. J. and Lepere, V. J., 1982, "Pool Boiling Heat Transfer From Enhanced Surfaces to Dielectric Fluids," ASME Journal of Heat Transfer, Vol. 104, pp. 292-299.

McBain, J. W., and Humphreys, C. W., 1932, "The Microtome Method of the Determination of the Absolute Amount of Adsorption," J. Chem. Phys., Vol. 36, pp. 300-311.

Miller, J. N., 1981, Volume Two Standards in Fluorescence Spectrometry, Chapman and Hall, London, pp. 44-67.

Semeria, R., 1962, "Quelques Resultats sur le Mecanisme de l'Ebullition," J. de l'Hydraulique de la Soc. Hydrotechnique de France, Vol. 7.

Stephan, K., 1963, "Influence of Oil on Heat Transfer of Boiling Refrigerant 12 and Refrigerant 22," XI Int. Congr. of Refrigeration, Vol. 1, pp. 369-380.

Webb, R.L., and Pais, C., 1992, "Nucleate Pool Boiling Data for Five Refrigerants on Plain, Integral-fin and Enhanced Tube Geometries," Int. J. Heat Mass Transfer, Vol. 35, No. 8, pp. 1893-1904.

Table 1 Conduction model choice

X_0= constant (all models) X_1= x X_2= y X_3= xy $X_4=x^2-y^2$ $X_5= y(3x^2-y^2)$ $X_6= x(3y^2-x^2)$ $X_7= x^4+y^4-6(x^2)y^2$ $X_8= yx^3-xy^3$	
Fluid	Most frequent models
R134a (File: 134apln.dat)	X_1,X_3,X_4 (83 of 144) 58 % X_1,X_2,X_3,X_4 (54 of 144) 38 %
R134a/DE589 (99.5/0.5) (File: 589pln5.dat)	X_1,X_2,X_3,X_4,X_6 (32 of 72) 44 % X_1,X_2,X_3,X_4 (14 of 72) 20 % X_1,X_3,X_4 (11 of 72) 15 %
R134a/DE589 (99/1) (File: 589pln1.dat)	X_1,X_3,X_4 (59 of 98) 60 % X_1,X_3,X_4,X_6 (29 of 98) 30 %
R134a/ DE589 (98/2) (File: 589pln2.dat)	X_1,X_3,X_4 (66 of 141) 47 % X_1,X_3,X_4,X_6 (31 of 141) 22 % X_1,X_2,X_4 (24 of 141) 17 %

Table 2 Pool boiling data

Pure R134a
File: 134apln.dat

ΔT_s (K)	q'' (W/m^2)
4.557	10041.4
4.364	8436.8
4.311	8227.6
4.308	8279.1
2.611	3423.0
4.406	8797.2
4.429	8959.4
3.597	5565.3
3.587	5578.7
3.561	5516.5
4.455	9517.4
2.586	4448.5
4.431	9793.0
4.401	9539.4
4.375	9352.1
7.028	73922.6
7.029	73196.7
7.021	73585.0
6.824	61446.9
6.845	64570.3
6.849	66446.8
6.650	54649.8
6.660	54579.2
6.676	54529.9
6.533	47396.7
6.523	47522.5
6.516	47720.3
6.247	35574.9
6.231	35693.4
6.226	35697.4
6.024	29734.6
6.035	29975.8
5.980	30428.2
5.978	29818.6
5.966	29290.0
5.665	21606.0
5.654	21634.3
5.648	21779.4
5.242	14056.2
5.230	14096.3
3.780	5843.6
3.733	5860.0
7.028	73922.6
7.029	73196.7
7.021	73585.0
6.824	61446.9
6.845	64570.3
6.849	66446.8
6.650	54649.8
6.660	54579.2
6.676	54529.9
6.533	47396.7
6.523	47522.5
6.516	47720.3
6.247	35574.9
6.231	35693.4
6.226	35697.4
6.024	29734.6
6.035	29975.8
5.980	30428.2
5.978	29818.6
5.966	29290.0
5.665	21606.0
5.654	21634.3
5.648	21779.4
5.242	14056.2
5.230	14096.3
3.780	5843.6
3.733	5860.0
6.964	74637.6
6.953	74417.3
6.956	75297.8
6.716	59173.0
6.716	59179.1
6.728	59056.2
6.703	57413.2
6.566	51163.0
6.603	52950.9
6.585	52187.4
6.437	45522.3
6.446	45546.7
6.465	45900.1
6.329	42922.7
6.366	43404.4
6.367	42746.1
6.259	37925.4
6.239	37862.2
6.225	37472.8
5.881	27216.4
5.870	27649.0
5.874	27593.5
5.704	22218.1
5.703	22395.4
5.699	22422.0
5.479	18290.6
5.470	18297.9
5.462	18355.0
4.697	9904.2
4.654	10264.4
4.644	10189.4
6.960	75309.8
6.948	75212.1
6.862	67431.9
6.851	67251.7
6.846	67108.2
6.700	61486.1
6.699	61588.0
6.680	61680.0
6.503	51515.8
6.505	51356.3
6.498	50898.3
6.414	46192.3
6.407	46618.4
6.400	46469.2
6.174	37587.4
6.174	37829.9
6.180	37937.8
5.890	28037.3
5.859	28090.4
5.840	28253.6
5.305	16946.5
5.302	16812.6
7.025	71589.6
7.027	72505.8
7.033	72986.3
6.818	62640.9
6.805	62456.8
6.801	62536.0
6.748	58529.1
6.739	57971.9
6.604	53891.6
6.627	53955.6
6.633	53882.7
6.511	48019.8
6.503	47628.0
6.491	47133.4
6.404	43432.4
6.394	43207.1
6.382	43208.4
6.255	39364.5
6.256	39507.9
6.264	39729.4
6.106	32179.1
6.090	31964.9
6.095	32169.4
5.940	27015.9
5.906	26558.2
5.872	26098.3
5.324	14898.6
5.320	14862.3

ΔT_s (K)	q'' (W/m²)
5.285	14744.1
5.001	12184.6
4.989	12140.5
7.026	69358.5
7.010	69098.3
7.000	68977.1
6.710	54797.8
6.716	54279.0
6.716	54365.9

R134a/DE589 (99.5/0.5)
File: 589pln5.dat

ΔT_s (K)	q'' (W/m²)
8.135	115792.8
8.150	116404.6
8.424	90755.8
8.422	90477.1
8.424	90077.3
8.094	71386.7
8.197	70388.5
8.197	69915.3
7.641	55685.5
7.618	55703.9
7.610	55808.8
7.011	41777.7
6.958	41567.3
6.940	41197.8
6.175	29217.0
6.161	29536.8
8.466	112110.9
8.454	112218.4
8.461	113065.2
7.955	87958.2
8.073	87557.6
8.154	86331.4
6.968	43403.3
6.972	44062.4
6.034	27166.5
6.052	27268.1
6.042	26948.1
4.713	11487.8
4.730	11572.3
4.760	11614.1
1.596	2786.9
8.388	111706.5
8.394	111698.3
8.373	111650.0
6.777	42008.5
6.817	42533.7
6.858	43201.7
5.946	26438.0

ΔT_s (K)	q'' (W/m²)
8.431	114899.1
6.593	42726.4
6.546	37613.2
5.869	26710.1
8.560	106942.9
8.389	101806.6
8.136	96172.8
7.913	81025.0
7.861	66613.5
7.609	59384.6
7.307	52775.4
4.730	11572.3
6.314	32699.3
5.388	19747.8
8.167	77346.8
8.016	72546.7
7.809	64845.3
4.128	9335.3
3.478	7053.9
3.042	5982.5
2.496	4752.8
8.882	112043.5
8.633	109038.6
7.973	86794.7
7.196	50004.7
7.078	46841.8
6.831	41468.4
6.509	34634.3
5.452	19784.2
4.504	11865.5
8.590	111722.6
6.818	42164.5
6.016	26944.5
4.888	14250.5
1.912	3404.6

R134a/DE589 (99/1)
File: 589pln1.dat

ΔT_s (K)	q'' (W/m²)
10.939	100016.1
10.755	80823.3
10.210	61838.5
8.715	33151.9
9.790	88712.2
9.916	83450.8
10.108	76137.2
9.939	70124.7
9.695	64981.9
9.544	61919.0
9.322	55292.6
9.073	48297.7
8.658	41691.3

ΔT_s (K)	q'' (W/m²)
8.373	36811.4
8.049	32615.0
7.475	26558.5
9.951	97818.8
9.777	79390.1
9.588	62321.6
8.350	36384.9
7.824	29278.0
7.373	24707.5
6.915	21209.8
6.537	18404.7
5.903	14533.7
5.170	11173.9
4.091	7580.5
9.595	77269.8
9.743	76651.4
9.637	68312.4
9.240	59474.3
8.950	51963.4
8.556	44534.2
8.289	38655.9
7.765	31411.4
9.552	76668.3
9.119	58089.9
8.686	48068.5
7.960	34347.8
8.038	35091.4
7.080	24068.4
9.814	67993.1
9.684	61932.1
9.455	63198.4
9.473	63064.3
8.566	45201.6
7.519	29255.3
7.004	23948.8
5.909	14180.4
4.404	8212.5
9.426	59391.8
9.387	60242.0
9.232	56766.7
8.855	49112.6
8.619	44298.4
8.021	36563.8
7.694	31548.6
6.962	22887.8
10.427	80329.2
10.523	80525.8
8.310	33854.1
7.976	30044.3
7.466	25506.0
6.780	20141.4
5.887	14266.6

ΔT_s (K)	q'' (W/m²)
8.838	41458.1
8.822	41704.6
8.826	41480.7
8.398	38094.4
8.379	37616.3
9.169	68668.1
8.738	54698.7
7.844	40726.1
7.851	40777.4
8.384	38349.0
8.382	38389.6
7.849	32759.2
8.169	39285.8
8.150	39265.7
6.793	19081.3
6.798	19230.2
6.819	19218.3
6.900	19865.6
6.966	20364.8
6.909	19660.6

R134a/ DE589 (98/2)
File: 589pln2.dat

ΔT_s (K)	q'' (W/m²)
7.186	18086.7
7.201	17863.7
7.240	17886.1
7.645	25700.9
7.687	25787.1
7.711	25612.1
6.397	11990.1
6.390	11967.4
6.372	11864.0
4.766	6431.0
4.757	6431.1
7.537	28984.5
7.556	28726.1
7.744	32903.0
7.519	27362.7
7.531	27235.3
7.186	21490.8
7.223	21860.5
7.207	22130.9
6.689	15668.1
6.685	15761.2
6.699	15848.0
5.284	8595.6
5.320	8569.3
7.827	45607.0
7.827	44786.2
7.824	44691.8
6.198	12273.6
6.232	12852.7
5.769	9990.1
5.729	9982.1
5.713	9842.0
5.195	7817.7
5.200	7834.6
4.815	6557.4
4.772	6542.7
3.555	4572.7
7.029	17636.4
7.054	17443.4
6.950	16418.2
6.932	16506.9
6.640	13984.9
6.659	14134.5
6.449	12715.8
6.459	12794.9
6.136	11281.3
6.081	11114.9
5.534	8879.4
5.539	8881.8
5.251	7893.8
5.248	7826.4
4.769	6669.0
7.361	22959.0
7.374	23056.1
6.581	14962.7
6.555	14612.0
4.051	5556.0
3.781	5356.2
3.734	5174.6
2.986	4097.4
2.979	4764.3
2.859	3878.7
2.845	3916.6
2.878	3888.7
7.462	36190.9
7.964	27087.5
7.953	27390.7
8.094	36326.3
8.072	35943.7
7.262	26252.0
7.451	27790.0
7.499	28420.9
6.582	17475.5
6.598	17754.5
6.465	15268.5
7.198	28948.5
10.633	141842.9
8.546	65681.4
8.382	58204.1
8.332	57389.2
8.057	50040.1
7.870	43679.6
7.689	35013.6
7.429	27463.8
8.900	74073.3
9.008	81913.5
8.936	85781.4
9.509	95938.3
9.642	95398.6
9.616	89179.1
9.625	88812.8
9.142	81875.2
9.077	82321.5
8.703	77453.5
8.690	78284.6
8.505	73557.1
8.483	73636.1
8.290	64416.9
7.999	55533.2
11.170	123502.0
11.261	122922.0
11.250	113726.7
11.065	110160.6
11.066	110923.8
10.215	78752.6
10.215	78752.6
10.193	77702.3
9.705	64873.0
9.599	65184.6
8.736	77911.8
8.651	79115.7
7.945	72996.3
7.993	70318.4
7.751	58493.4
7.822	58444.0
7.661	44684.8
11.080	128082.3
11.025	129260.0
10.583	114869.1
10.556	119695.9
9.614	94875.4
9.596	96202.7
9.094	75539.6
9.109	74972.0
8.723	56773.2
10.709	128565.2
10.696	128282.8
10.200	109381.7
10.191	109414.6
9.702	90982.4
9.515	90650.8
9.152	75616.6

9.117	74819.3
8.779	65973.7
8.778	65882.6
8.478	54770.4
10.309	112886.2

10.083	111072.4
9.056	70277.2
9.026	70993.5
8.629	56290.8
10.279	113188.5

10.223	114290.3
9.878	103487.9
9.780	103377.9

Table 3 Number of test days and data points

Fluid (% mass)	Number of days	Number of data points
R134a	6	144
R134a/DE589 (99.5/0.5)	9	72
R134a/DE589 (99/1)	13	85
R134a/DE589 (98/2)	15	141

Table 4 Constants for cubic boiling curve fits for plain copper surface

$$\Delta T_s = A_0 + A_1 q'' + A_2 q''^2 + A_3 q''^3$$

ΔT_s in Kelvin and q'' in W/m^2

Fluid	A_o	A_1	A_2	A_3
R134a				
$3 K \leq \Delta T_s \leq 6 K$	1.13413	5.40212x10^{-4}	-2.23805x10^{-8}	3.26420x10^{-13}
$6 K \leq \Delta T_s \leq 7 K$	5.08549	3.62330x10^{-5}	-1.57067x10^{-10}	2.86909x10^{-16}
R134a/DE589 (99.5/0.5)	2.93162	1.46223x10^{-4}	-1.41993x10^{-9}	4.99448x10^{-15}
$3.6 K \leq \Delta T_s \leq 8.5 K$				
R134aa/DE589 (99/1)	3.44201	2.16478x10^{-4}	-2.88547x10^{-9}	1.53149x10^{-14}
$11 K \leq \Delta T_s \leq 21.5 K$				
R134aa/DE589 (98/2)				
$3 K \leq \Delta T_s \leq 7 K$	-1.76162	1.53377x10^{-3}	-1.01205x10^{-7}	2.41953x10^{-12}
$7 K \leq \Delta T_s \leq 11 K$	6.91642	1.57640x10^{-5}	1.97728x10^{-10}	-5.36523x10^{-16}

Table 5 Residual standard deviation of ΔT_s from the mean

Fluid	U (K)
R134a	
$3 K \leq \Delta T_s \leq 6 K$	0.1
$6 K \leq \Delta T_s \leq 7 K$	0.04
R134a/DE589 (99.5/0.5)	0.19
$3.6 K \leq \Delta T_s \leq 8.5 K$	
R134a/DE589 (99/1)	0.26
$11 K \leq \Delta T_s \leq 21.5 K$	
R134aa/DE589 (98/2)	
$3 K \leq \Delta T_s \leq 7 K$	0.14
$7 K \leq \Delta T_s \leq 11 K$	0.39

Table 6 Average magnitude of 95 % multi-use confidence interval for mean T_w-T_s(K)

Fluid	U (K)
R134a	
3 K ≤ ΔT_s ≤ 7 K	0.09
7 K ≤ ΔT_s ≤ 11 K	0.03
R134aa/DE589 (99.5/0.5) 3.6 K ≤ ΔT_s ≤ 8.5 K	0.15
R134aa/DE589 (99/1) 11 K ≤ ΔT_s ≤ 21.5 K	0.19
R134aa/DE589 (98/2)	
3 K ≤ ΔT_s ≤ 7 K	0.15
7 K ≤ ΔT_s ≤ 11 K	0.26

Table 7 Extrapolated raw fluorescence intensity measurements

R134a/DE589 (99.5/0.5)
File: 5%589e.feh

F_m	s_F	$T_s(K)$	$\Delta T_s(K)$
3.81766	.015	277.45	7.00
2.75722	.015	276.96	7.65
1.52147	.016	276.97	7.26
3.49533	.015	277.59	5.80
2.76672	.015	276.52	8.14
2.55146	.016	276.67	7.05
6.14011	.016	277.47	6.78
6.43490	.014	277.41	1.39
-.40767	.015	277.19	6.89
2.28226	.015	277.65	6.78
2.16089	.016	276.94	7.07
.42281	.015	277.44	6.51
1.68678	.015	277.91	6.18
.22995	.014	277.27	6.89
2.45310	.015	276.91	7.26
.73493	.014	276.92	6.82
4.40181	.015	277.31	4.99
-2.91581	.015	277.55	6.55
1.38028	.015	277.18	6.71
2.97000	.015	277.43	5.85
.77982	.403	277.50	3.87
1.14627	.016	277.37	6.51
.85105	.015	276.96	7.33
.38793	.015	277.55	5.79
.57564	.015	277.49	5.46
.27591	.014	277.47	4.29
2.99769	.015	278.02	2.21

R134a/DE589 (99/1)
File: 1%589e.feh

F_m	s_F	$T_s(K)$	$\Delta T_s(K)$
9.74034	.015	274.52	9.24
14.94173	.016	276.67	7.89
13.50717	.015	276.81	8.40
21.08775	.014	276.35	8.15
17.89300	.015	277.04	6.82
21.52693	.015	276.85	6.82
13.35840	.015	276.89	6.58
19.11047	.015	276.94	6.62
17.18062	.015	276.98	5.53
14.54255	.015	277.09	5.26
13.47910	.015	276.64	8.00
13.70351	.015	276.64	8.67
10.98960	.015	276.92	6.65
11.26164	.015	276.82	6.79
15.35011	.016	277.08	6.10
15.54583	.197	276.92	6.72
13.45906	.189	276.96	6.41
12.65105	.229	277.19	4.85
16.22508	.016	277.23	5.28
12.04902	.015	277.60	5.11
14.99398	.015	277.43	4.27
10.46799	.015	277.09	5.35
14.82547	.015	276.92	6.85
84.99503	.016	276.22	9.15
9.74037	.015	276.22	8.72
14.22953	.015	275.93	6.29

R134a/DE589 (98/2)
File: 2%589e.feh

F_m	s_F	$T_s(K)$	$\Delta T_s(K)$
36.57075	.015	276.62	8.05
38.70269	.015	276.85	8.04
37.02534	.014	276.30	8.47
30.87293	.016	276.39	7.83
38.21518	.015	276.33	7.09
34.19254	.015	276.15	8.00
32.86617	.015	276.54	6.68
36.32044	.015	277.00	5.41
31.01951	.015	275.80	8.07
44.05413	.014	275.44	7.80
24.14667	.015	276.33	8.03
40.46933	.016	276.00	7.85
37.95875	.016	275.88	7.70
39.22655	.015	275.05	7.95
39.88002	.015	276.64	7.11
43.01120	.015	275.16	8.01
42.83043	.015	276.38	8.26
37.87346	.015	276.08	7.47
39.08668	.015	276.39	6.50
59.61776	.015	275.08	8.26
66.83802	.015	275.63	7.51
50.01350	.015	275.96	7.35
54.30996	.015	275.39	8.31
48.76741	.015	275.84	7.25
64.69425	.015	275.92	6.52

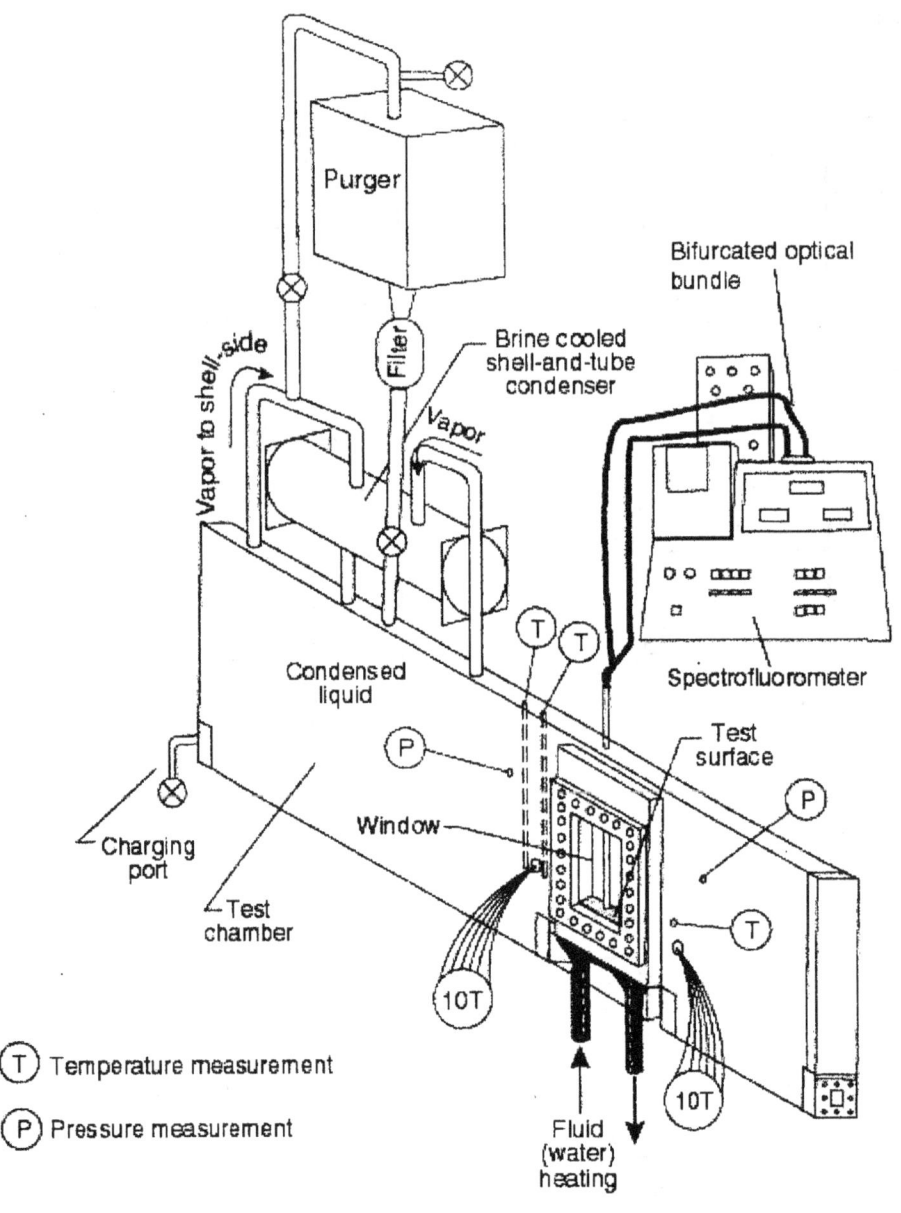

Fig. 1 Schematic of test apparatus

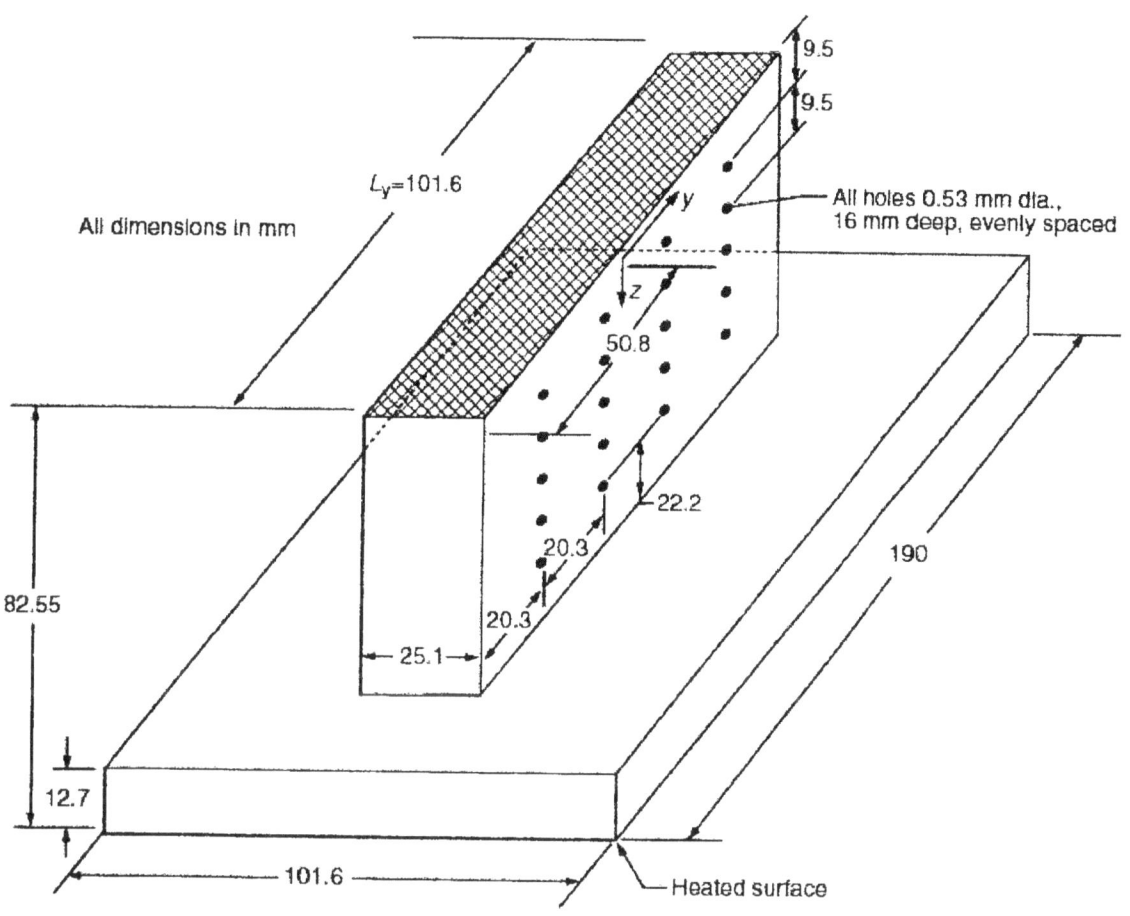

Fig. 2 OFHC copper flat test plate with cross-hatched surface and thermocouple coordinate system

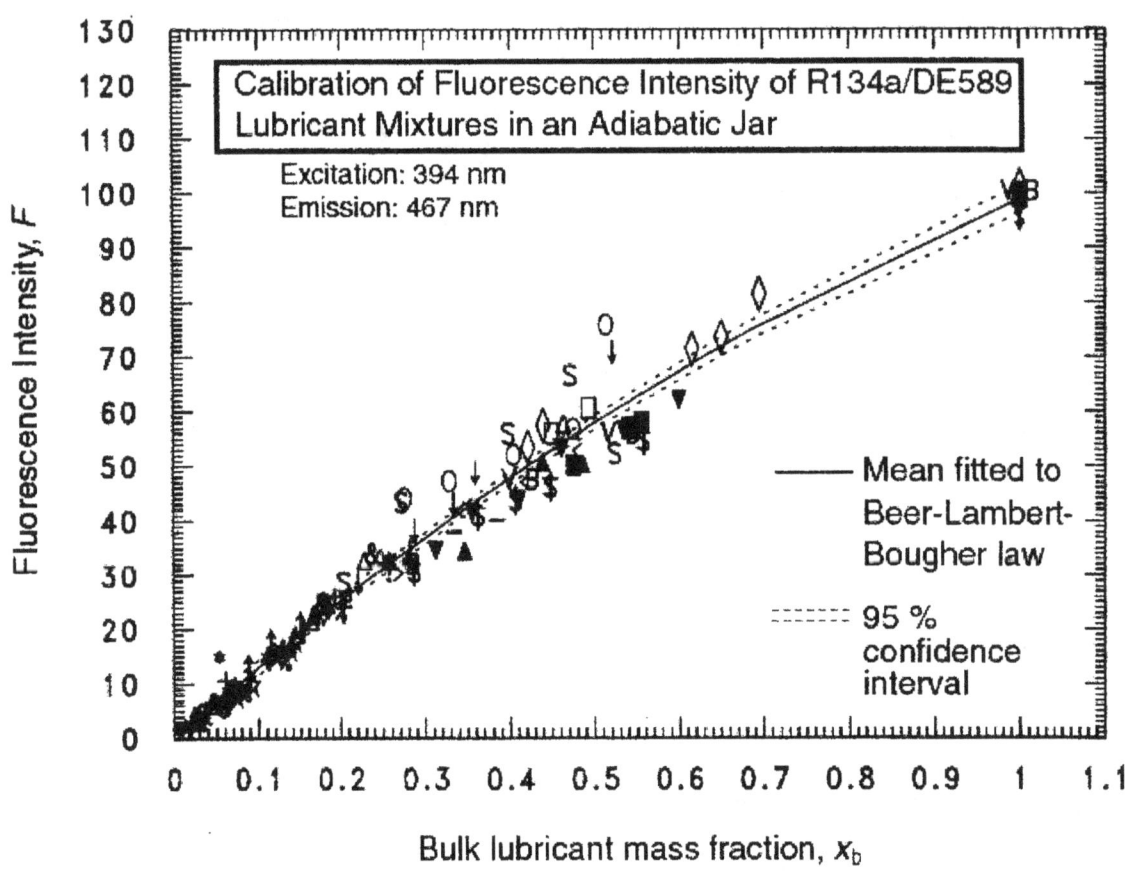

Fig. 3 Fluorescence calibration with F = 100 for 100 % DE589 jar

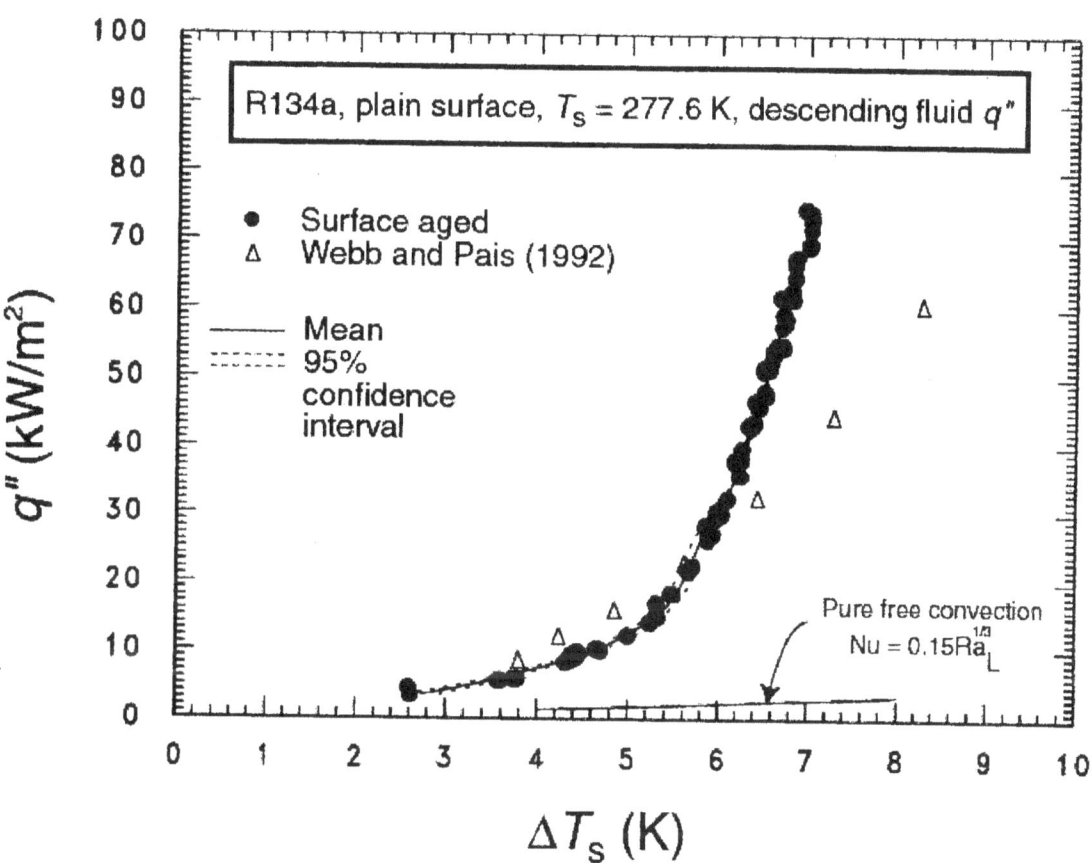

Fig. 4 Pure R134a boiling curve for plain surface

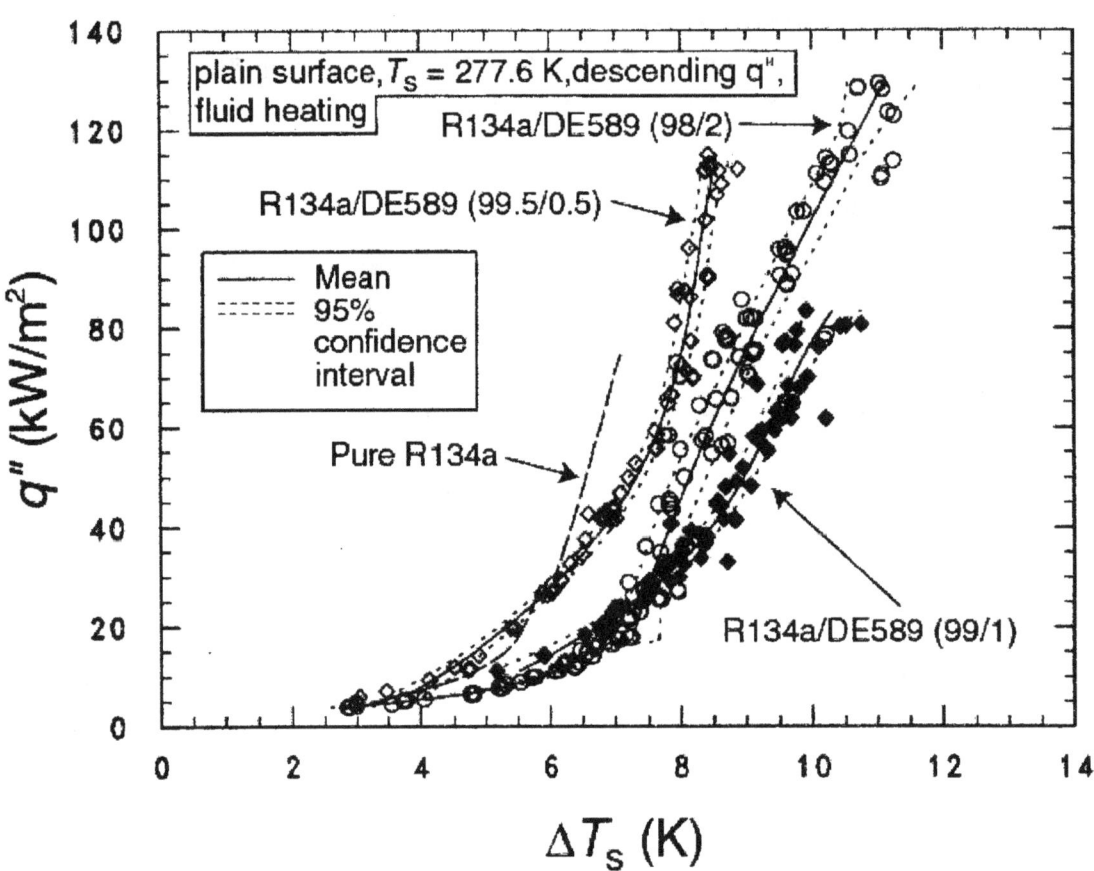

Fig. 5 Three R134a/DE589 mixture boiling curves for plain surface

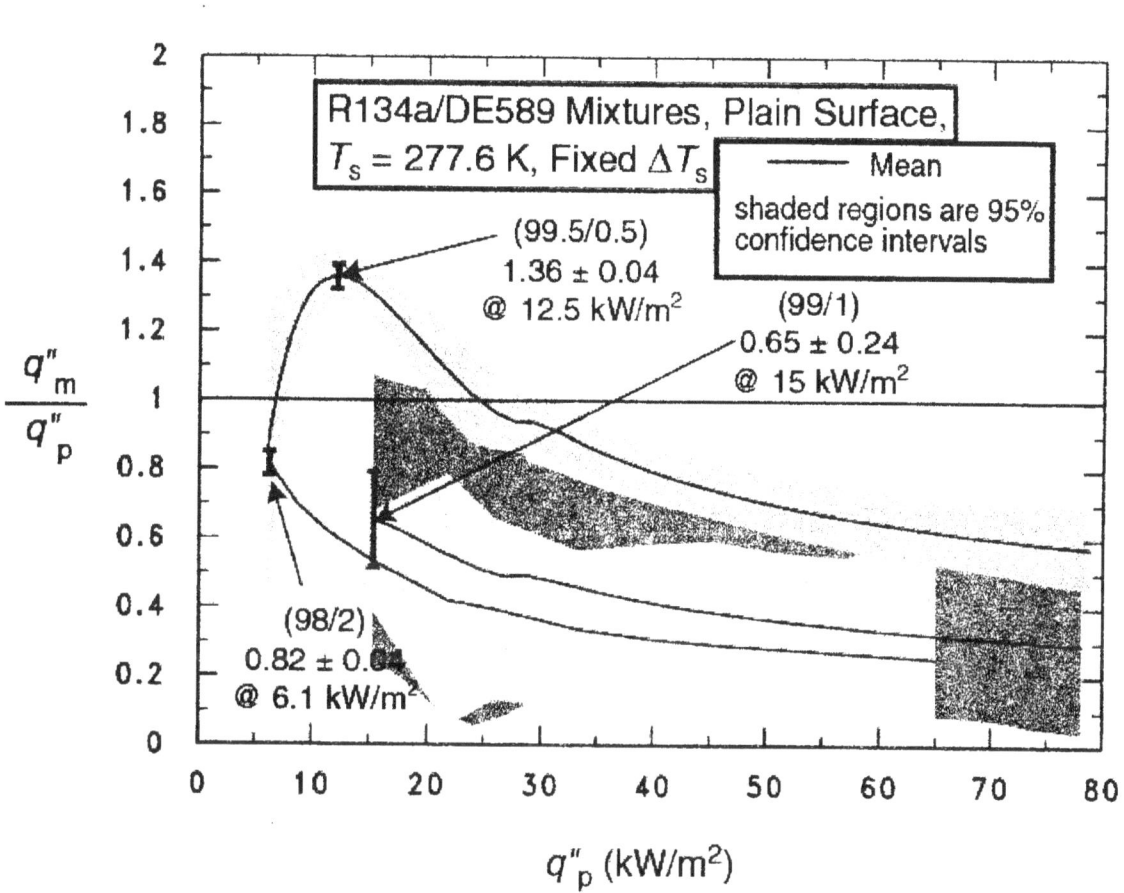

Fig. 6 Three R134a/DE589 mixture heat fluxes relative to pure R134a for a plain surface

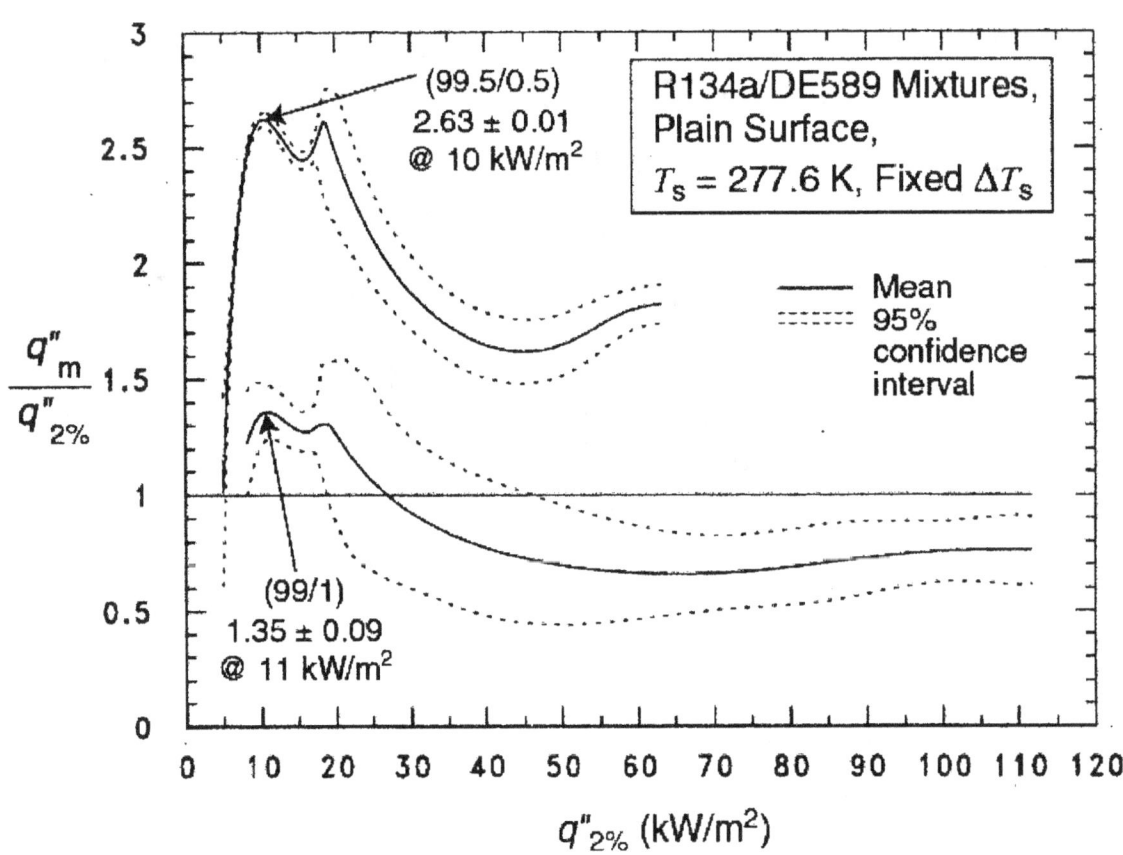

Fig. 7 Two R134a/DE589 mixture heat fluxes relative to R134a/DE589 (98/2) for a plain surface

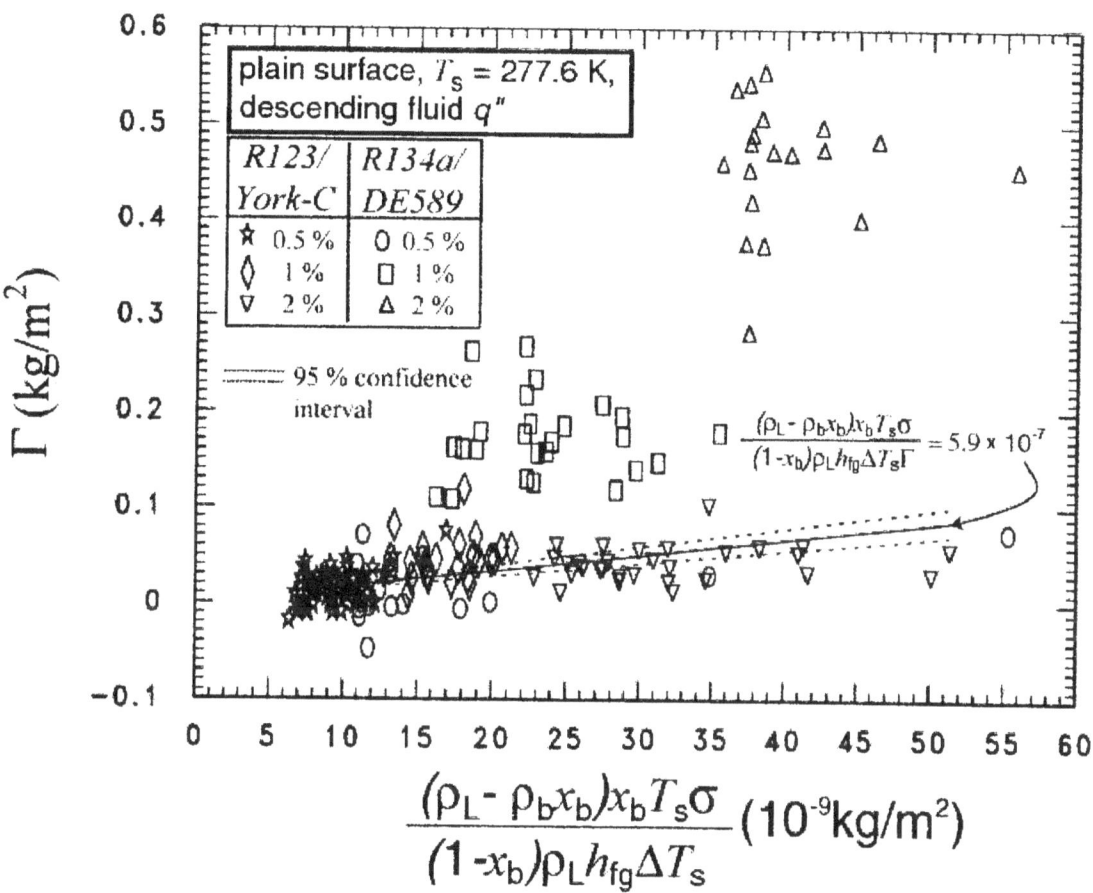

Fig. 8 Lubricant excess surface density for six refrigerant/lubricant mixtures as a function of non-dimensional excess surface density as defined in Kedzierski (2003)

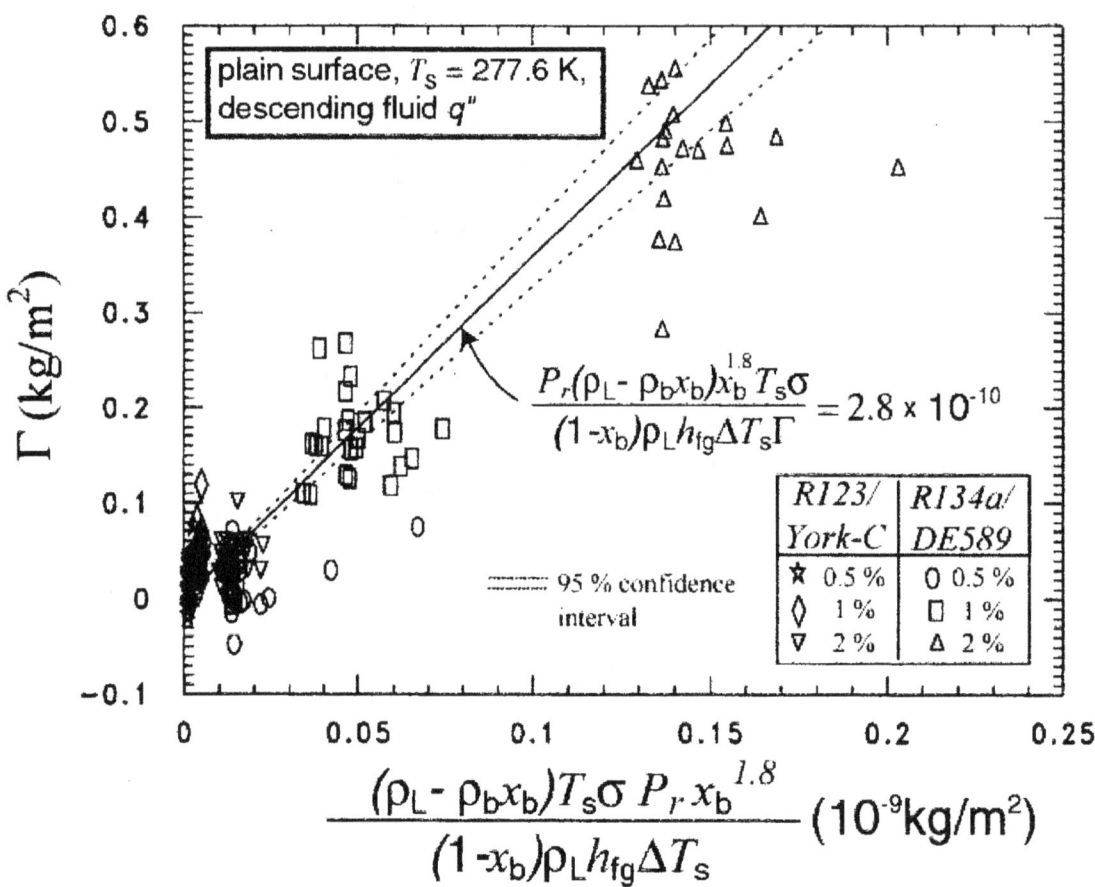

Fig. 9 Lubricant excess surface density for six refrigerant/lubricant mixtures as a function of pressure corrected non-dimensional excess surface density

APPENDIX A

Figure A.1 shows the relative (percent) uncertainty of the heat flux ($U_{q''}$) as a function of the heat flux. Figure A.2 shows the uncertainty of the wall temperature as a function of heat flux. The uncertainties shown in Figs. A.1 and A.2 are "within-run uncertainties." These do not include the uncertainties due to "between-run uncertainties" or differences observed between tests taken on different days. The "within-run uncertainties" include only the random effects and uncertainties evident from one particular test. All other uncertainties reported here are "between-run uncertainties" which include all random effects such as surface past history or seeding. "Within-run uncertainties" are given only in Figs. A.1 and A.2.

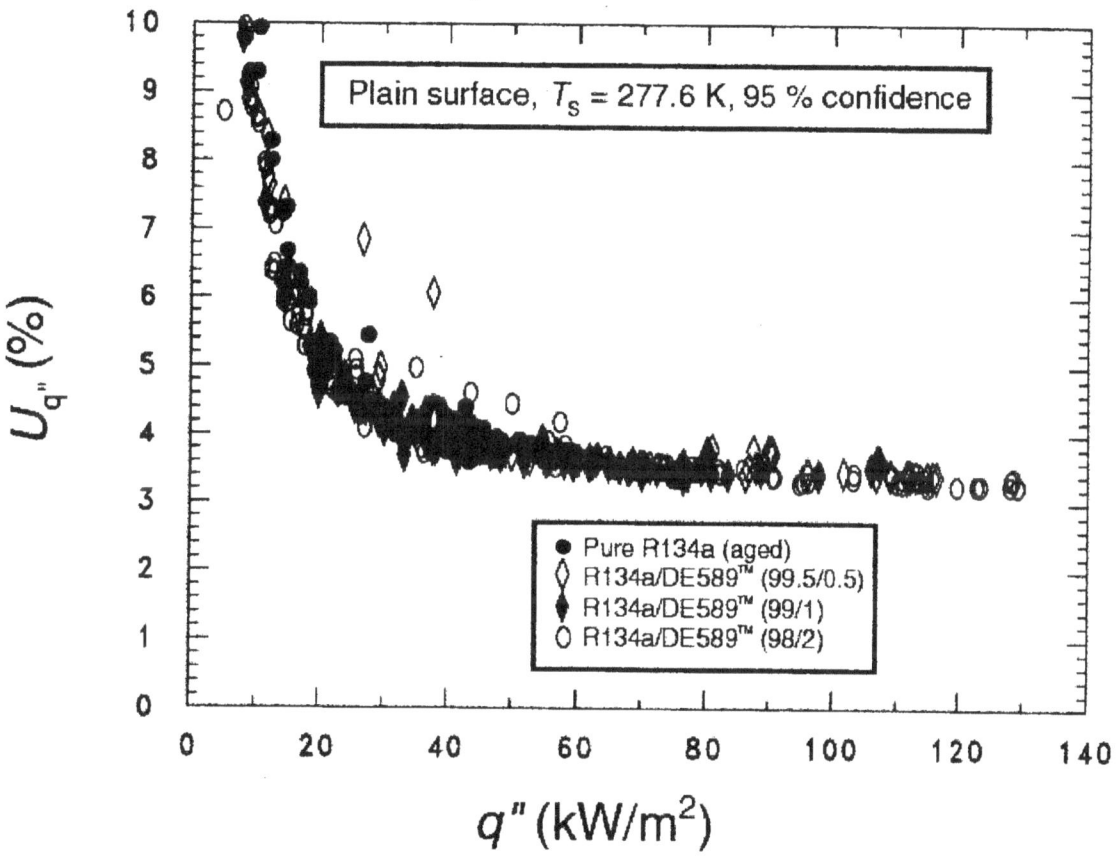

Fig. A.1 Uncertainty in the heat flux at surface for 95% confidence

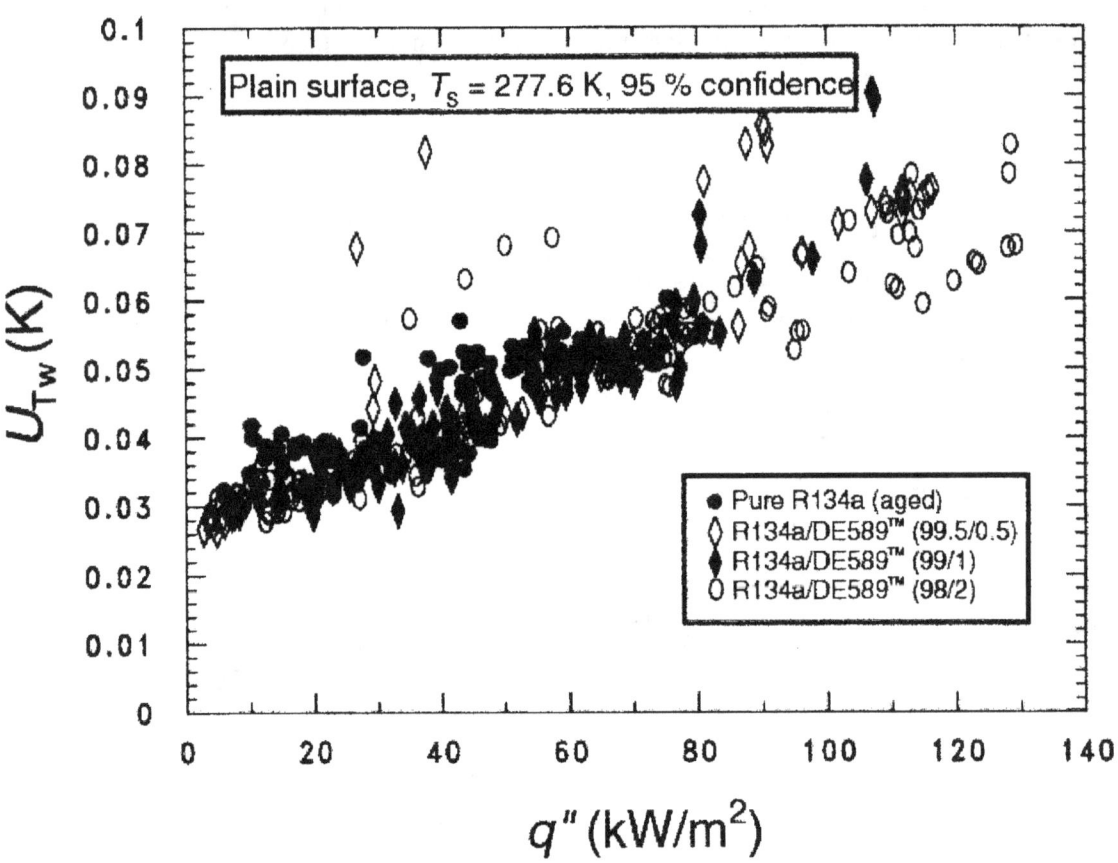

Fig. A.2 Uncertainty in the temperature of the surface for 95% confidence

APPENDIX B

Figure B.1 shows the measured emission and excitation spectra for pure DE589 in a cuvette. The test sample was placed directly in the sample chamber of the right angle spectrofluorometer. The excitation wavelength that produced the maximum fluorescence emission was iteratively found by scanning through both excitation and emission wavelengths. The excitation and emission wavelengths for DE589 that produced the largest intensities were located at 394 nm and 467 nm, respectively.

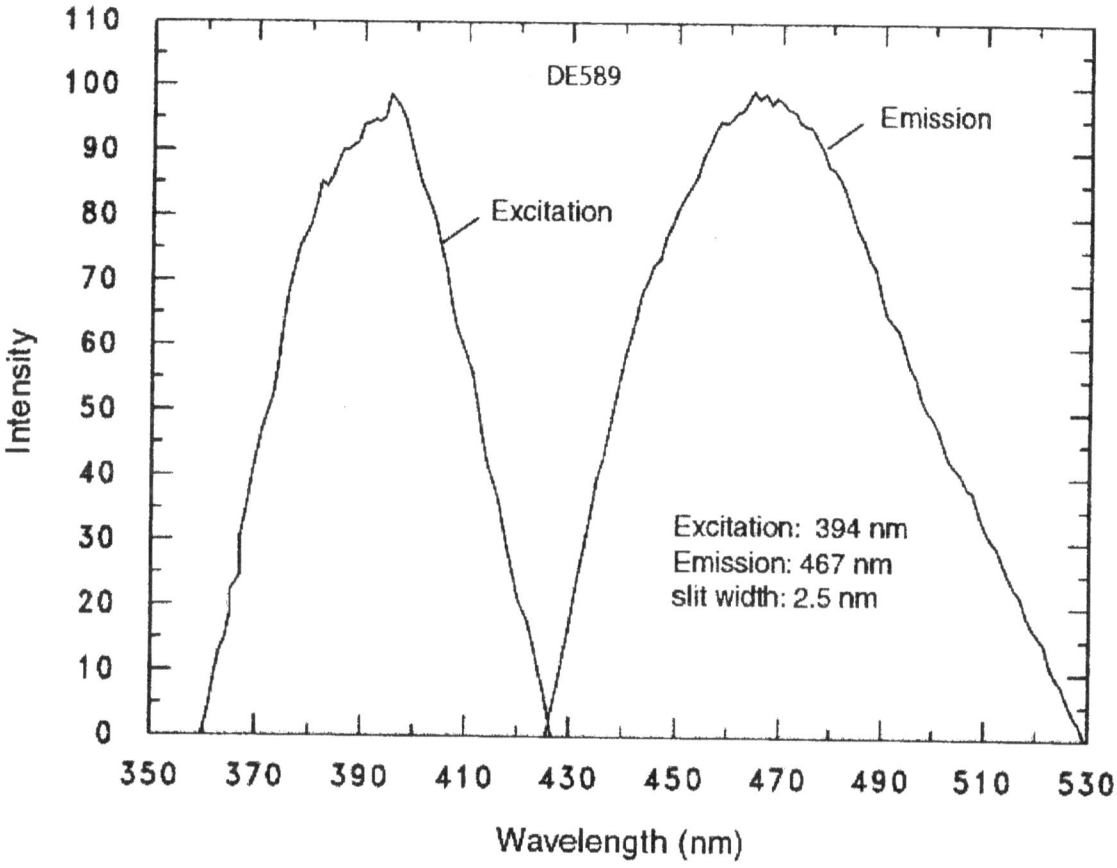

Fig. B.1 Fluorescence spectra for pure DE589 lubricant

APPENDIX C

Figure C.1 shows the sample chamber of the spectrofluorometer with two 10 nm bandwidth bandpass interference filters placed at the exit of the excitation monochromator and before the entrance to the excitation monochromator. The peak transmitted wavelength for the excitation filter was 394 nm. The peak transmitted wavelength for the emission filter was 467 nm.

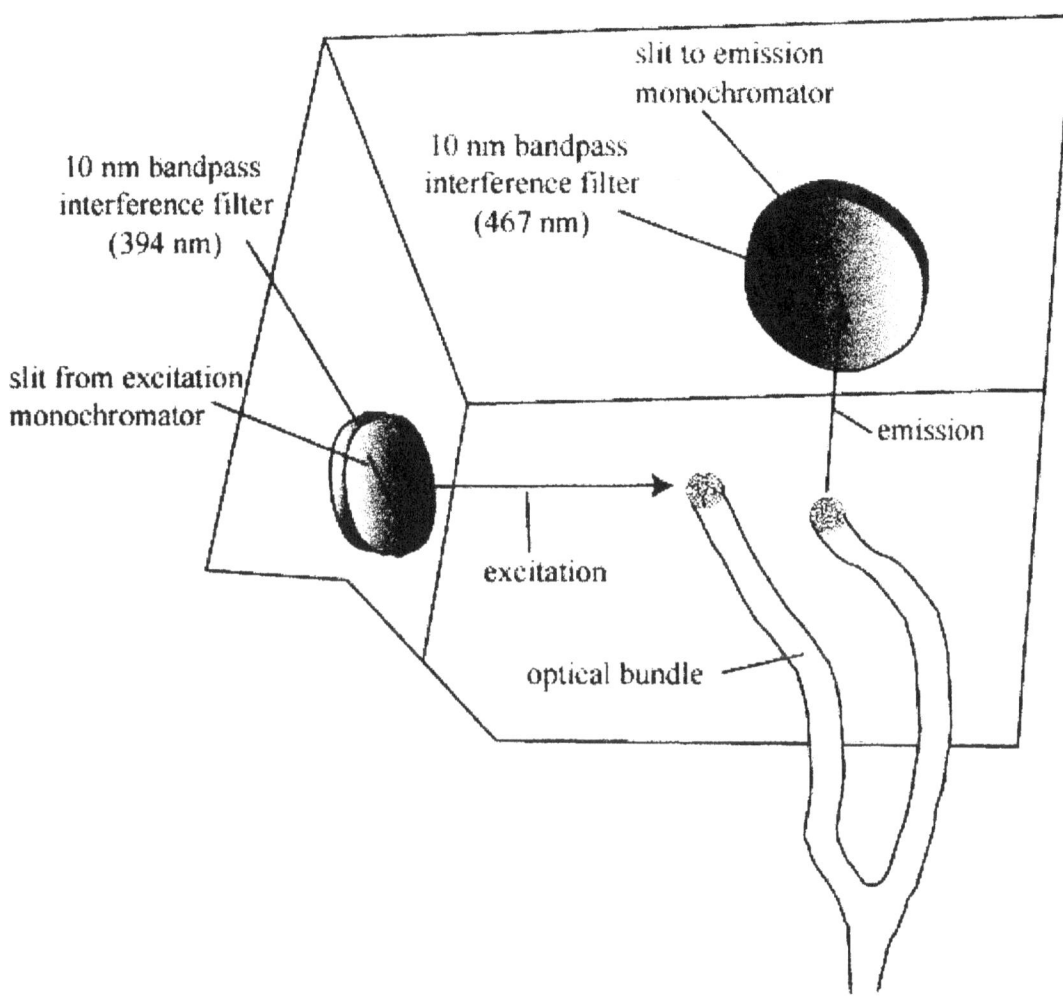

Fig. C.1 Spectrofluorometer sample chamber with interference filters

APPENDIX D

Kedzierski (2002a) describes the method for calibrating the emission intensity measured with the spectrofluorometer and the bifurcated optical bundle as shown in Fig. 2 against the bulk lubricant mass fraction. Three glass vessels were each fitted with a glass tube of the same type that was used in the test chamber. Two jars were used to set the lower (0) and upper (100) limits of the intensity signal on the spectrofluorometer. A jar that contained only pure R134a was used to zero the intensity. Because light intensities are additive, the zeroing ensured that the reflected excitation wave and other effects were not attributed to fluorescence. A second jar that contained a 0.5 mass fraction[3] liquid mixture of R134a and DE589 was used to set the intensity on the spectrofluorometer to 100. The third jar was used to measure and record the intensity of prepared refrigerant/lubricant mixtures of various concentrations. The third jar was initially charged with approximately 20 g of lubricant and then evacuated for approximately 10 s. Evacuation of the jar and the sample prevented fluorescence quenching by oxygen (Guilbault, 1967). The jar was then charged with approximately 20 g of pure R134a to give approximately a 0.5 mass fraction. Calibration measurements proceeded by successively diluting the mixture with approximately 2 g increments of pure R134a.

A single calibration run consisted of measurements for concentrations beginning with a 0.5 mass fraction and diluting to a lubricant mass fraction of 0.05 or less. Prior to each emission intensity measurement for the variable jar, the zero and 100 limits for the emission intensity of the spectrofluorometer were set with the pure R134a jar and the 50/50 jar, respectively. All emission measurements were made at a wavelength of 467 nm with an excitation wavelength of 394 nm. Although, the calibration data was taken at room temperature, both the pure refrigerant jar and the 50/50 jar were maintained with a constant temperature bath to within approximately 1 K of the temperature of the saturated refrigerant in the boiling rig during heat transfer/fluorescence measurements to account for the temperature effect on fluorescence (Miller, 1981).

The fluorescence measurements that were used to fit Eq. (1) are given in Table D.1. The mass of the lubricant and refrigerant charged are presented along with the refrigerant quality and the lubricant mass fraction.

---

[3] Liquid composition assuming that some refrigerant but no lubricant is in the vapor phase.

Table D.1 Measured fluorescent intensity in calibration jar (DE589cal.dat)

F	s_F	T(K)	m_L (kg)	m_r (kg)	x_{df}	x_b
75.85716	0.44571	297.600	0.010004	0.013052	0.272561	0.51306436096
56.76649	0.49763	297.630	0.010004	0.014581	0.241351	0.47489151798
52.03908	0.48121	297.640	0.010004	0.018197	0.188046	0.40372685557
47.19302	0.47420	297.650	0.010004	0.023725	0.137936	0.32846878614
44.08457	0.41048	297.660	0.010004	0.029520	0.105550	0.27477342579
60.62425	0.18228	296.390	0.010011	0.013727	0.248499	0.49250113983
55.96784	0.14350	297.390	0.010011	0.015731	0.220104	0.44933606112
57.22449	0.13073	297.540	0.010011	0.015081	0.231802	0.46355374173
32.21911	0.11943	297.570	0.010011	0.036789	0.079101	0.22809329118
24.42234	0.11886	297.580	0.010011	0.047111	0.055843	0.18371776463
20.96322	0.10977	297.590	0.010011	0.053412	0.046067	0.16421576028
18.54149	0.10854	297.610	0.010011	0.062092	0.035852	0.14326631046
17.89467	0.08915	297.470	0.010011	0.061805	0.036003	0.14385509707
16.17688	0.09599	296.480	0.010011	0.070285	0.027628	0.12776602142
9.21856	0.09096	297.523	0.010011	0.130114	0.002907	0.07163674660
6.75032	0.07514	297.540	0.010011	0.158753	-.002506	0.05918001426
101.3592	0.16764	297.400	---	0.00	0.0	1.0
81.47487	0.16537	297.420	0.009989	0.008057	0.455936	0.69500649382
73.57123	0.17004	297.460	0.009989	0.009029	0.404438	0.65005773049
71.43201	0.17125	297.480	0.009989	0.009871	0.367857	0.61550762057
57.44873	0.11113	297.480	0.009989	0.016193	0.213681	0.43962054172
53.45592	0.13274	297.480	0.009989	0.017157	0.200155	0.42126455034
23.91303	0.10146	297.550	0.009989	0.048481	0.053467	0.17876482910
15.34111	0.09522	297.610	0.009989	0.071284	0.027737	0.12597133162
14.53040	0.09003	297.620	0.009989	0.080664	0.021358	0.11232408408
8.68224	0.09679	297.180	0.009989	0.122750	0.004693	0.07558091976
7.23732	0.08828	297.190	0.009989	0.145661	-.000257	0.06416058381
57.90660	0.15885	297.310	0.010005	0.011593	0.307563	0.55483380695
56.95831	0.15754	297.310	0.010005	0.012056	0.294718	0.54058062857
50.14035	0.14703	297.310	0.010005	0.014523	0.240087	0.47549528810
49.93558	0.14979	297.570	0.010005	0.014113	0.249800	0.48585482134
50.10550	0.15848	297.570	0.010005	0.016249	0.213397	0.43907543319
34.04970	0.14071	297.530	0.010005	0.022159	0.149068	0.34666385635
32.20074	0.11937	297.540	0.010005	0.032000	0.094921	0.25675228890
24.17035	0.11287	297.540	0.010005	0.046927	0.056107	0.18425734805
22.08364	0.11758	297.560	0.010005	0.052907	0.046729	0.16553689208
16.31673	0.12922	297.640	0.010005	0.064729	0.033316	0.13785266883
9.75148	0.15746	297.724	0.010005	0.098864	0.012449	0.09295019427
7.71859	0.13831	297.740	0.010005	0.125581	0.003999	0.07406515284
62.60534	0.17898	298.120	0.010067	0.010411	0.353908	0.59945954682
57.05283	0.17948	298.520	0.010067	0.012442	0.295153	0.53443508009
53.79970	0.17305	298.580	0.010067	0.015371	0.233992	0.46091573089
44.21058	0.17610	298.600	0.010067	0.018010	0.195710	0.41002326379
42.16451	0.15267	298.610	0.010067	0.021716	0.157565	0.35495516143
35.16792	0.17237	298.640	0.010067	0.025478	0.130266	0.31238680932
34.70880	0.14663	298.330	0.010067	0.028950	0.110258	0.28100461596
30.71385	0.15008	298.360	0.010067	0.032266	0.096151	0.25661100864
25.71732	0.14558	298.500	0.010067	0.040508	0.071209	0.21109032108
19.61072	0.13129	298.520	0.010067	0.057788	0.041569	0.15380548163
11.47472	0.12639	298.500	0.010067	0.104424	0.010491	0.08877775422
5.27890	0.14777	298.180	0.010067	0.161717	-.003083	0.05843305206
99.54967	0.26746	297.050	---	0.000010	0.0	1.0
54.63259	0.23192	297.950	0.010009	0.011571	0.314215	0.55778413383
46.13726	0.22119	297.350	0.010009	0.015776	0.219139	0.44827389452
43.71872	0.20113	297.150	0.010009	0.018006	0.187528	0.40623627615
40.99505	0.17556	297.250	0.010009	0.020877	0.158538	0.36295788616
30.76662	0.20967	297.950	0.010009	0.028142	0.113014	0.28621221454
23.12728	0.13518	298.150	0.010009	0.042842	0.065186	0.19994690685
14.01769	0.11214	297.750	0.010009	0.068329	0.030218	0.13122557452
8.17443	0.08681	297.350	0.010009	0.110352	0.008250	0.08379197105

F	s_F	$T(K)$	m_l (kg)	m_t (kg)	x_{qf}	x_b
5.61356	0.08803	298.050	0.010009	0.160594	-.002876	0.05850996792
99.09577	0.23040	297.880	------	0.000010	0.0	1.0
34.06634	0.16009	297.877	0.004818	0.018938	0.190216	0.23906287814
24.75055	0.14355	297.870	0.004818	0.025867	0.131893	0.17665606773
15.63938	0.09889	297.870	0.004818	0.038767	0.078885	0.11888415642
6.40308	0.09220	297.868	0.004818	0.095272	0.015845	0.04887381785
4.46973	0.08066	297.865	0.004818	0.133789	0.003394	0.03487441469
3.84302	0.09271	297.862	0.004818	0.168496	-.002948	0.02771981201
3.29833	0.09004	297.861	0.004818	0.168496	-.002948	0.02771981674
14.16790	0.19355	297.783	0.000997	0.021100	0.173582	0.05408358345
8.62255	0.16818	297.820	0.000997	0.018070	0.207501	0.06508913013
5.30071	0.17673	297.825	0.000997	0.025897	0.136539	0.04268334023
2.54289	0.14717	297.783	0.000997	0.041761	0.074185	0.02513869882
0.94732	0.15633	297.838	0.000997	0.055768	0.048767	0.01844745717
1.56621	0.14854	297.849	0.000997	0.076198	0.028360	0.01328731030
0.92825	0.16891	297.852	0.000997	0.171027	-.002549	0.00578105281
97.88289	0.27562	297.912	0.000010	0.010011	0.401029	1.0
47.85656	0.17069	297.910	0.010011	0.015817	0.222217	0.44865844937
40.23001	0.12267	297.969	0.010011	0.019429	0.176115	0.38476822420
38.15573	0.14847	297.973	0.010011	0.023273	0.142502	0.33406125572
33.06446	0.12130	297.974	0.010011	0.028848	0.109652	0.28045338072
32.54288	0.12059	297.986	0.010011	0.028848	0.109691	0.28046222455
23.09483	0.11237	297.972	0.010011	0.042014	0.066668	0.20337639166
20.97473	0.09827	297.971	0.010011	0.059520	0.038971	0.14894782861
17.74524	0.10015	297.966	0.010011	0.079275	0.022404	0.11439848591
13.10391	0.08894	297.964	0.010011	0.105449	0.010018	0.08750601796
95.26597	2.53350	297.822	0.009995	0.000010	0.0	1.0
70.90511	0.29556	297.830	0.009995	0.012783	0.280837	0.52089681116
48.94271	0.22039	297.820	0.009000	0.019478	0.176582	0.35944583450
43.10279	0.22771	297.810	0.009995	0.023266	0.141884	0.33361231928
38.19288	0.17451	297.800	0.009995	0.028077	0.112852	0.28636144529
28.79727	0.18503	297.800	0.009995	0.038200	0.075702	0.22062459217
24.06867	0.15742	297.800	0.009995	0.046708	0.056932	0.18494256735
15.06273	0.14950	297.803	0.009995	0.081583	0.020910	0.11121355704
100.4096	0.35859	297.350	0.010006	0.000010	0.0	1.0
56.01876	0.22804	297.250	0.010006	0.012399	0.285299	0.53032766617
47.55285	0.26009	297.750	0.010006	0.017702	0.194704	0.41242551057
32.18174	0.24033	297.250	0.010006	0.030072	0.101860	0.27032382529
26.33442	0.16012	297.650	0.010006	0.044748	0.060351	0.19222567869
14.57155	0.12445	273.150	0.010006	0.080154	0.010375	0.11201365038
10.77543	0.09755	297.450	0.010006	0.154630	-.001836	0.06067185790
96.17478	0.27886	297.650	0.010169	0.000010	0.0	1.0
52.74501	0.19415	297.550	0.010169	0.014536	0.241216	0.47969952447
47.01308	0.22342	297.250	0.010169	0.017559	0.193263	0.41788423283
41.88180	0.19226	297.350	0.010169	0.022798	0.143113	0.34234098108
30.44302	0.15504	297.450	0.010169	0.030876	0.098920	0.26767039552
25.24081	0.15471	297.950	0.010169	0.042662	0.065070	0.20315658306
15.79468	0.11778	297.550	0.010169	0.065542	0.032400	0.13818929972
11.46506	0.13069	297.650	0.010169	0.103185	0.010714	0.09059363380
100.0438	0.21784	297.508	0.009826	0.000010	0.0	1.0
99.37194	0.28411	297.514	0.010208	0.000010	0.0	1.0
52.25241	0.17547	297.570	0.010208	0.012842	0.276676	0.52356972724
66.57565	0.24124	297.603	0.010208	0.014958	0.233932	0.47113457780
55.97153	0.17565	297.450	0.010208	0.018828	0.179438	0.39785575385
43.36567	0.18187	297.850	0.010208	0.030593	0.101237	0.27074146307
28.69437	0.18370	297.850	0.010208	0.042639	0.064899	0.20383486796

APPENDIX E

This appendix presents the measurements and the correlation of the DE589 lubricant liquid density (ρ_b). The density of the liquid lubricant was measured as a function of temperature with a glass pycnometer. The pycnometer was factory instrumented with a glass mercury thermometer with a range of 14°C to 38°C in 0.2° graduations, accurate to within ±0.2 K. The pycnometer was filled with distilled water and its volume was calculated from the known density of water. The volume was found over five trails to be 9.84 ml with a standard uncertainty of 0.01 ml.

The pycnometer containing DE589 was cooled in an ice bath and then removed from the bath and allowed to warm on the balance to room temperature over approximately one hour. The standard uncertainty of the balance was approximately 1 mg. The outside of the pycnometer was wiped clean before each measurement to remove the lubricant that was expelled through the pipette due to volume expansion with temperature increase. Condensed water from the air was also wiped from the outside of the pycnometer for some data points.

The Biot number for the warming pycnometer was estimated to be approximately 0.5, which is greater than the recommended limit of 0.1 (Incropera and Dewitt, 1985) for a uniform temperature in fluid. It is difficult to estimate the error introduced in the measurements due to temperature gradients that existed in the lubricant. The first measurement (0 °C) and the last measurements (ambient) were considered to have uniform temperatures. Removal of data for temperatures less than 16.5 °C (with the exception of 0 °C) resulted in regression residuals that were independent of temperature, which suggests that the error due to temperature gradients in the liquid was negligible.

Table E.1 shows the recorded mass and temperature for two days. Equation (E.1) gives the fit of the liquid lubricant density (ρ_L) in kg/m^3 versus temperature (T) in Kelvin:

$$\rho_L = 1205.8 - 0.779018T \tag{E.1}$$

The expanded uncertainty of the fit was approximately ± 0.7 kg/m^3 for 95 % confidence. Figure E.1 shows a plot of the density versus temperature measurements and the equation (E.1) fit with 95 % confidence intervals.

Table E.1 DE589 liquid density measurements

T (°C)	m (g)	ρ_L (kg/m³)
.0	9.770	993.3
14.0	9.654	981.5
14.5	9.649	981.0
15.0	9.643	980.4
16.5	9.641	980.2
17.0	9.636	979.7
17.5	9.634	979.5
18.0	9.631	979.2
18.5	9.628	978.9
19.0	9.626	978.7
19.5	9.623	978.3
20.0	9.617	977.7
20.5	9.614	977.4
21.0	9.610	977.0
21.5	9.606	976.6
22.0	9.602	976.2
22.5	9.598	975.8
23.0	9.594	975.4
23.5	9.590	975.0
.0	9.770	993.3
14.0	9.653	981.4
14.5	9.650	981.1
15.0	9.645	980.6
15.5	9.641	980.2
16.0	9.638	979.9
16.5	9.635	979.6
17.0	9.631	979.2
17.5	9.628	978.9
18.0	9.624	978.4
18.5	9.620	978.0
19.0	9.616	977.6
19.5	9.614	977.4
20.0	9.611	977.1
20.5	9.608	976.8
21.0	9.604	976.4
21.5	9.600	976.0
22.0	9.598	975.8
22.5	9.594	975.4
23.0	9.590	975.0
23.5	9.587	974.7
.0	9.762	992.5
14.0	9.649	981.0
14.5	9.644	980.5
15.0	9.641	980.2
15.5	9.638	979.9
16.0	9.636	979.7
16.5	9.634	979.5
17.0	9.630	979.1
17.5	9.628	978.9
18.0	9.627	978.8
18.5	9.624	978.4
19.0	9.622	978.2
19.5	9.618	977.8
20.0	9.612	977.2
20.5	9.608	976.8
21.0	9.604	976.4
21.5	9.600	976.0
22.0	9.598	975.8
22.5	9.594	975.4
23.0	9.590	975.0
23.5	9.587	974.7
.0	9.771	993.4
14.0	9.654	981.5
14.5	9.652	981.3
15.0	9.650	981.1
15.5	9.648	980.9
16.0	9.644	980.5
16.5	9.642	980.3
17.0	9.640	980.1
17.5	9.636	979.7
18.0	9.634	979.5
18.5	9.630	979.1
19.0	9.627	978.8
19.5	9.621	978.1
20.0	9.618	977.8
20.5	9.614	977.4
21.0	9.610	977.0
21.5	9.607	976.7
22.0	9.602	976.2
22.5	9.596	975.6
23.0	9.592	975.2
23.5	9.590	975.0
.0	9.770	993.3
14.0	9.652	981.3
14.5	9.649	981.0
15.0	9.645	980.6
15.5	9.644	980.5
16.0	9.642	980.3
16.5	9.642	980.3
17.0	9.640	980.1
17.5	9.634	979.5
18.0	9.627	978.8
18.5	9.624	978.4
19.0	9.622	978.2
19.5	9.615	977.5
20.0	9.611	977.1
20.5	9.607	976.7
21.0	9.605	976.5
21.5	9.602	976.2
22.0	9.598	975.8
22.5	9.594	975.4
23.0	9.591	975.1
23.5	9.590	975.0

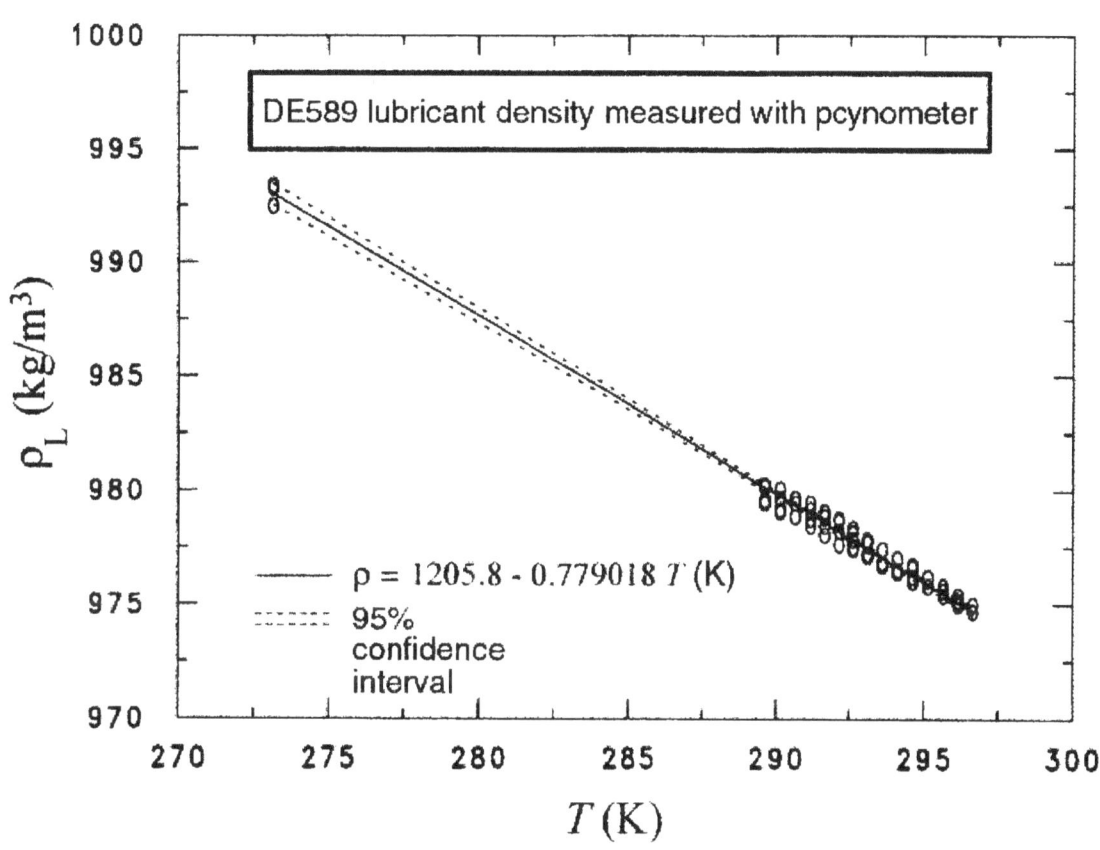

Fig. E.1 Fit of lubricant density measurements with 95 % confidence intervals

APPENDIX F

This appendix outlines the development of the expression for the excess surface density while allowing for a difference in temperature between the lubricant excess layer and the bulk fluid.

A simplified form of the Beer-Lambert-Bougher law valid for dilute solutions (Amadeo et al., 1971) can be used to represent the ratio of the fluorescence intensity of the excess layer (F_e) to that of the bulk fluid (F_b) under the probe:

$$\frac{F_e}{F_b} = \frac{I_{oe} \varepsilon_e x_e \rho_e l_e \Phi_e}{I_{ob} \varepsilon_b x_b \rho_b l_b \Phi_b} \tag{F.1}$$

where Φ is the quantum efficiency of the fluorescence, ε is the extinction coefficient, l is the path length, and I_o is the intensity of the incident radiation.

Equation (F.1) can be corrected, so that it is applicable for non-dilute solutions, by multiplying by a correction term (Kedzierski, 2001a):

$$\frac{F_{e,T_b}}{F_{b,T_b}} = \left(\frac{I_{oe} \varepsilon_e x_e \rho_e l_e \Phi_e}{I_{ob} \varepsilon_b x_b \rho_b l_b \Phi_b} \right)_{T_b} \left(\frac{1 + 1.165 \frac{\varepsilon_b}{M_L} x_b \rho_b l_b}{1 + 1.165 \frac{\varepsilon_e}{M_L} x_e \rho_e l_e} \right)_{T_b} \tag{F.2}$$

Here both F_e and F_b are evaluated at the bulk fluid temperature, T_b.

For heat transfer at the wall-excess layer interface, the excess layer and the bulk fluid will be at different temperatures. The effect of temperature on fluorescence of the excess layer can be represented as:

$$\frac{F_{T_e}}{F_{T_b}} = e^{\beta(T_e - T_b)} \tag{F.3}$$

The derivation of Eq. (F.3) is given in Appendix G.

The fluorescence of the lubricant excess layer (F_e) evaluated at the average excess layer temperature (T_e) can be obtained by substituting Eq. (F.2) into Eq. (F.3):

$$F_{e,T_e} = F_{b,T_b} \left(\frac{I_{oe} \varepsilon_e x_e \rho_e l_e \Phi_e}{I_{ob} \varepsilon_b x_b \rho_b l_b \Phi_b} \right)_{T_b} \left(\frac{1 + 1.165 \frac{\varepsilon_b}{M_L} x_b \rho_b l_b}{1 + 1.165 \frac{\varepsilon_e}{M_L} x_e \rho_e l_e} \right)_{T_b} e^{\beta(T_e - T_b)} \tag{F.4}$$

$$F_{e,T_e} = F_m - F_b \tag{F.5}$$

Set the above two expressions for F_{e,T_e} equal and solve for $x_e \rho_e l_e$

$$x_e \rho_{e,T_b} l_e = \frac{\left(\dfrac{F_m}{F_b} - 1\right)}{\dfrac{I_{oe}\varepsilon_e \Phi_e}{I_{ob}\varepsilon_b \Phi_b x_b \rho_b l_b}\left(1 + 1.165\dfrac{\varepsilon_b}{M_L} x_b \rho_b l_b\right)e^{\beta(T_e - T_b)} - 1.165\dfrac{\varepsilon_e}{M_L}\left(\dfrac{F_m}{F_b} - 1\right)} \quad \text{(F.6)}$$

The mass of lubricant on the surface can be expressed in terms of the density of the excess layer evaluated at the its own temperature ($\rho_{e,Te}$):

$$x_e \rho_{e,T_e} l_e = x_e \rho_{e,T_b} l_e \frac{\rho_{e,T_e}}{\rho_{e,T_b}} \quad \text{(F.7)}$$

Substitution of Eq. (F.6) into Eq. (F.7) gives:

$$x_e \rho_{e,T_e} l_e = \frac{\dfrac{\rho_{e,T_e}}{\rho_{e,T_b}}\left(\dfrac{F_m}{F_b} - 1\right)}{\dfrac{I_{oe}\varepsilon_e \Phi_e}{I_{ob}\varepsilon_b \Phi_b x_b \rho_b l_b}\left(1 + 1.165\dfrac{\varepsilon_b}{M_L} x_b \rho_b l_b\right)e^{\beta(T_e - T_b)} - 1.165\dfrac{\varepsilon_e}{M_L}\left(\dfrac{F_m}{F_b} - 1\right)} \quad \text{(F.8)}$$

The excess surface density (Γ) as defined in this study is:

$$\Gamma = x_e \rho_{e,T_e} l_e - x_b \rho_b l_e \quad \text{(F.9)}$$

Substitute Eq. (F.8) into Eq. (F.9) to get:

$$\Gamma = x_b \rho_b \left\{\frac{\dfrac{\rho_{e,T_e}}{\rho_{e,T_b}}\left(\dfrac{F_m}{F_b} - 1\right)}{\dfrac{I_{oe}\varepsilon_e \Phi_e}{I_{ob}\varepsilon_b \Phi_b l_b}\left(1 + 1.165\dfrac{\varepsilon_b}{M_L} x_b \rho_b l_b\right)e^{\beta(T_e - T_b)} - 1.165\dfrac{\varepsilon_e}{M_L}\left(\dfrac{F_m}{F_b} - 1\right)x_b \rho_b} - l_e\right\} \quad \text{(F.10)}$$

Solving Eq. (F.8) for the minimum excess layer thickness by setting $x_e = 1$, and $\rho_e = \rho_L$ yields:

$$l_{e,min} = \frac{\dfrac{1}{\rho_{e,T_b}}\left(\dfrac{F_m}{F_b} - 1\right)}{\dfrac{I_{oe}\varepsilon_e \Phi_e}{I_{ob}\varepsilon_b \Phi_b x_b \rho_b l_b}\left(1 + 1.165\dfrac{\varepsilon_b}{M_L} x_b \rho_b l_b\right)e^{\beta(T_e - T_b)} - 1.165\dfrac{\varepsilon_e}{M_L}\left(\dfrac{F_m}{F_b} - 1\right)} \quad \text{(F.11)}$$

Substituting Eq. (F.11) into Eq. (F.10) for l_e and continuing to apply $\rho_e = \rho_L$ gives:

$$\Gamma = \frac{x_b \rho_b \left(\dfrac{\rho_{L,T_e}}{\rho_{L,T_b}} - \dfrac{x_b \rho_b}{\rho_{L,T_b}}\right)\left(\dfrac{F_m}{F_b} - 1\right)}{\dfrac{I_{oe}\varepsilon_e \Phi_e}{I_{ob}\varepsilon_b \Phi_b l_b}\left(1 + 1.165\dfrac{\varepsilon_b}{M_L} x_b \rho_b l_b\right)e^{\beta(T_e - T_b)} - 1.165\dfrac{\varepsilon_e}{M_L}\left(\dfrac{F_m}{F_b} - 1\right)x_b \rho_b} \quad \text{(F.12)}$$

Considering that ε_e, ε_b, Φ_e, and Φ_b are evaluated at T_b as shown in Eq. (F.4) and that these coefficients are for the same lubricant:

$$\varepsilon_e = \varepsilon_b = \varepsilon$$
$$\Phi_e = \Phi_b \qquad \text{(F.13)}$$

Use of Eq. (F.13) to simplify Eq. (F.12) gives:

$$\Gamma = \frac{x_b \rho_b \left(\dfrac{\rho_{L,T_e}}{\rho_{L,T_b}} - \dfrac{\rho_b x_b}{\rho_{L,T_b}} \right)\left(\dfrac{F_m}{F_c} - 1 \right)}{\dfrac{I_{oe}}{I_{ob}}\left(1 + 1.165 \dfrac{\varepsilon}{M_L} x_b \rho_b l_b \right) \dfrac{e^{\beta(T_e - T_b)}}{l_b} - 1.165 \dfrac{\varepsilon}{M_L} x_b \rho_b \left(\dfrac{F_m}{F_c} - 1 \right)} \qquad \text{(F.14)}$$

where all properties are evaluated at T_b except $\rho_{L,Te}$.

APPENDIX G

This appendix presents the measurements and the methodology that were used to determine the coefficient of temperature dependence of the fluorescence (β) for the DE589 lubricant. A spectrofluorometer was used with capped cuvettes to measure the fluorescence intensity of eight different mixtures of DE589 and R123. The R123 was used because it, like R134a, is a nonfluorescent fluid, and because it has a vapor pressure near atmospheric, which was necessary considering the cuvette was not a pressure vessel. The zero of the spectrofluorometer was set with the photomultiplier-tube shutter closed. The maximum range (100 %) of the spectrofluorometer was set with the shutter open with a pure DE589 cuvette at room temperature in the sample chamber. All of the measurements and settings were made with the excitation set to 394 nm and the emission read at 467 nm. Each prepared mixture was cooled to approximately 280 K and then allowed to heat to room temperature while it was in the sample chamber. One sample was also heated to approximately 308 K and allowed to cool to room temperature. The temperature of the mixture was measured with a radiation pyrometer while the cuvette remained in the sample chamber of the spectrofluorometer. A temperature measurement was taken immediately before the fluorescence intensity measurement.

Figure G.1 shows measured fluorescence intensity versus temperature for eight different mixtures of DE589 and R123. A common observed characteristic of the F versus T plot for each fluid is that the slope decreases as the fluorescent intensity decreases:

$$\frac{\left(\frac{\partial F}{\partial T}\right)_T}{F(T)} = C_1 \tag{G.1}$$

Equation (G.1) is valid regardless of how the magnitude of F was obtained. In other words, fluorescence intensities with an average intensity of 20 % would exhibit the same slope as given by Eq. (G.1) whether the intensity was obtained with a more dilute solution and the 100 % of the spectrofluorometer set with a 50/50 mixture or a less dilute solution and the 100 % of the spectrofluorometer set with pure lubricant. Equation (G.1) is independent of composition and spectrofluorometer settings and only depends on the magnitude of the intensity. This is an important characteristic because it allows Eq. (G.1) to be applied as a correction for temperature regardless of spectrofluorometer setting and sample composition.

Separation of the variables and integration of Eq. (G.1) yields:

$$F = C_3 e^{C_1 T} \tag{G.2}$$

Figure G.1 shows that Eq. (G.2) was used to regress the eight data sets of measured F and T. The solid line is the mean of the regression. The dashed lines to either side of the mean represent the lower and upper 95 % simultaneous (multiple-use) confidence intervals for the mean. Table G.1 shows the regression statistics for each line. The statistics verify Eq. (G.1) in that the regressed values for C_1 for each data set are comparable to one another. An uniformly weighted average of all of the C_1 coefficients yields: -0.01 $K^{-1} \pm 0.008$ K^{-1}. A t-value-weighted average of the three C_1 constants with the smallest uncertainties yields: -0.01 $K^{-1} \pm 0.001$ K^{-1}.

For the same fluid and spectrofluorometer zero and maximum settings, the ratio of the intensity at T_e to that at T_b can be derived from Eq. (G.2) as:

$$\frac{F_{T_e}}{F_{T_b}} = e^{C_1(T_e-T_b)} = e^{\beta(T_e-T_b)} \tag{G.3}$$

where the constant C_1 is equivalent to the coefficient of temperature dependence of the fluorescence (β).

Table G.1 Fluorescence temperature dependence regression statistics

Data file name	$C_1=\beta$ (K^{-1})	% standard dev. ($100\ \sigma_{C_1}/C_1$)	t-value
Fvst1.dat	-0.0139	14	7.4
Fvst2.dat	-0.0085	21	4.6
Fvst3.dat	-0.012	10	10
Fvst4.dat	-0.0076	25	3.9
Fvst5.dat	-0.01	10	10
Fvst6a.dat	-0.0086	5.8	16
Fvst7dat	-0.014	26	3.7
Fvst8.at	-0.0176	28	3.8

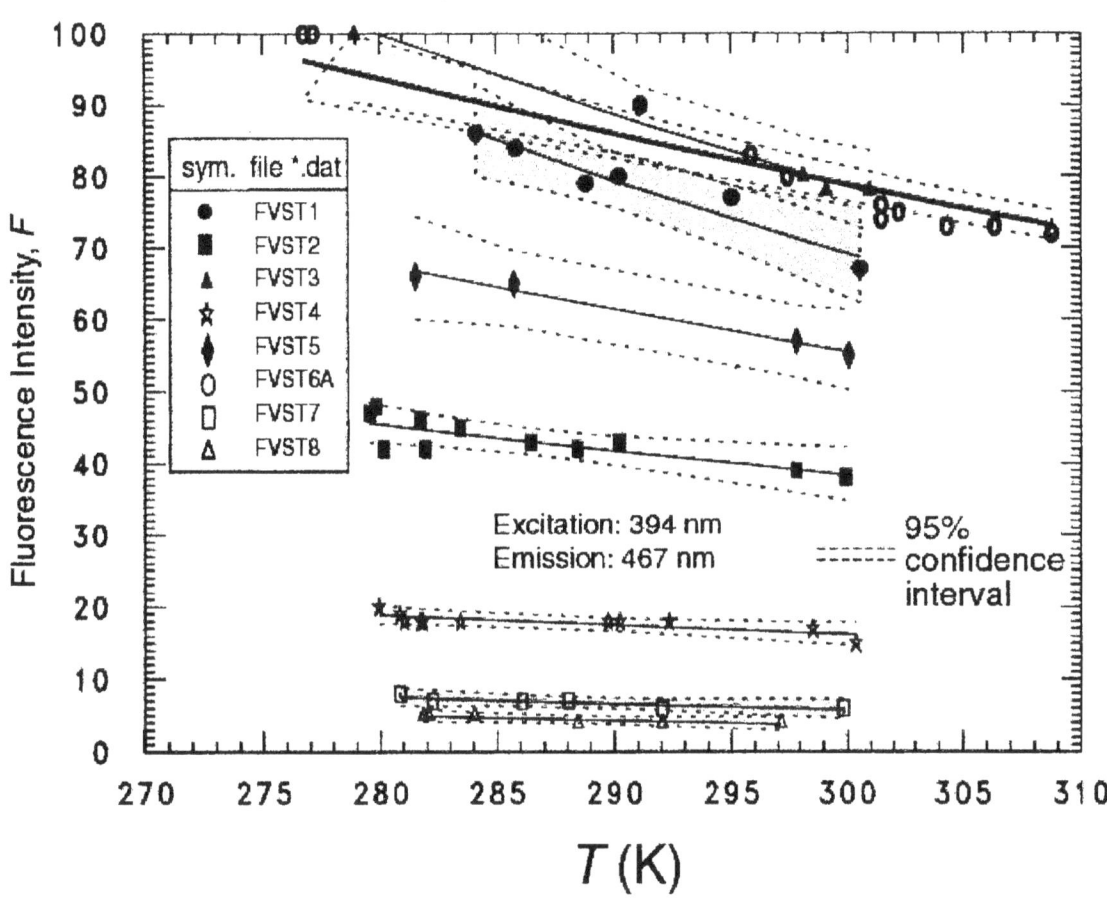

Fig. G.1 Measured fluorescence intensity versus temperature for eight different mixtures of DE589 and R123

APPENDIX H

This appendix gives a sample calculation of the ratio of the absorption of the incident excitation in the bulk to that in the excess layer (I_{oe}/I_{ob}). The absorption of a R134a/DE589 (92.9/7.1) mass fraction mixture was measured in an ultraviolet/visible absorption spectrometer and is shown in Fig. H.1. Figure H.1 shows two runs with a single but correct cuvette and two runs with both a sample and a reference cell. Two beams are analyzed in the spectrometer. The first beam is incident the sample: a cuvette with the R134a/DE589 mixture, while the second beam is incident the reference cell: a cuvette with pure R134a. The absorption is recorded as the difference between the absorption of the sample and that of the reference cell. The difference in absorption measured in this manner gives the absorption of only the lubricant in the mixture. The runs made without the reference cell were corrected by manually subtracting the absorption of an empty quartz cuvette from the single cuvette measurements. The absorption calculated in this manner is the absorption for the refrigerant and the lubricant at room temperature (297.6 K). Figure H.1 shows that there is little difference between the two measurement methods for wavelengths above 280 nm. A small difference between the measurements with and without a reference cell occurs for wavelengths below 280 nm which may be due to absorption by R134a or another unidentified effect. The results suggest that the refrigerant does not absorb for the excitation wavelengths (greater than 280 nm) used in this study.

The Beer-Lambert Law (Amadeo et al., 1971) relates the absorbance (A) to the ratio of the incident light intensity (I_o) to the transmitted light intensity (I_t), i.e.,

$$A = \varepsilon c l = \log_{10} \frac{I_o}{I_t} \tag{H.1}$$

The ratio of the absorbance of a R134a/DE589 (92.9/7.1) mass fraction mixture ($A_{7.1}$) to that of a R134a/DE589 (98/2.0) mass fraction mixture ($A_{2.0}$) at the pool boiling test temperature (277.6 K) can be calculated as:

$$\frac{A_{7.1}}{A_{2.0}} = \frac{c_{7.1}}{c_{2.0}} = \frac{x_{m_7} \rho_{m_7}}{x_{m_2} \rho_{m_2}} = \frac{0.071(1259 \text{kg/m}^3)}{0.020(1274 \text{kg/m}^3)} = 3.5 \tag{H.2}$$

Correction of the absorption measurements for temperature can be done with use of the relationship between the fluorescence and absorbed light (I_a) intensities (Amadeo et al., 1971):

$$F = I_a \Phi \tag{H.3}$$

If it is assumed that, π, the quantum efficiency of fluorescence is independent of temperature for the present temperature range, Eq. (G.3) can be used to express the functional dependence of absorbed light with respect to temperature:

$$\frac{I_{a,T_1}}{I_{a,T_2}} = \frac{A_{T_1}}{A_{T_2}} = e^{\beta(T_1-T_2)} \tag{H.4}$$

The absorption for a 2.0 % mass mixture at 380 nm at 277.6 K can be calculated from the absorption for the 7.1 % mass mixture (1.43) and Eq. (H.2) as:

$$A_{2.0,277.6K} = A_{7,297.6K} \frac{A_{7,277.6K}}{A_{7,297.6K}} \frac{A_{2.0}}{A_7} = 1.43 e^{-0.01(277.6-297.6)} \frac{1}{3.5} = 0.044 \tag{H.5}$$

The product of ε and c for the 2.0 % mass mixture at 394 nm can be calculated from Eq. (H.1):

$$\varepsilon c = \frac{A_{2.0}}{l_c} = \frac{0.044}{10\,\text{mm}} = 0.0044\,\text{mm}^{-1} \tag{H.6}$$

where the absorption length (l_c) is the internal thickness of the cuvette that was used in the absorption spectrometer.

The absorption ratio I_o/I_t can be calculated from the εc product and Eq. (H.1) with the absorption length l equal to the distance between the bottom of the glass tube and the copper surface, i.e.,

$$\frac{I_o}{I_t} = 10^{\varepsilon c l} = 10^{0.0044\,\text{mm}^{-1} \cdot 1.46\,\text{mm}} = 1.015 \tag{H.7}$$

For the optical bundle, the incident intensity (I_o) is equal to the incident intensity for the bulk fluid (I_{ob}). Similarly, the transmitted intensity of Eq. (H.7) is equal to the incident intensity for the excess layer (I_{oe}). Accordingly, the ratio of the absorption of the incident excitation in the bulk to that in the excess layer (I_{oe}/I_{ob}) for the 2.0 % mass mixture at 394 nm excitation is:

$$\frac{I_{oe}}{I_{ob}} = \frac{1}{1.015} = 0.985 \tag{H.6}$$

The same calculation for I_{oe}/I_{ob} was done for the 0.5 % and the 1.0 % mass mixtures for a 394 nm excitation and was found to be 0.996 and 0.993, respectively.

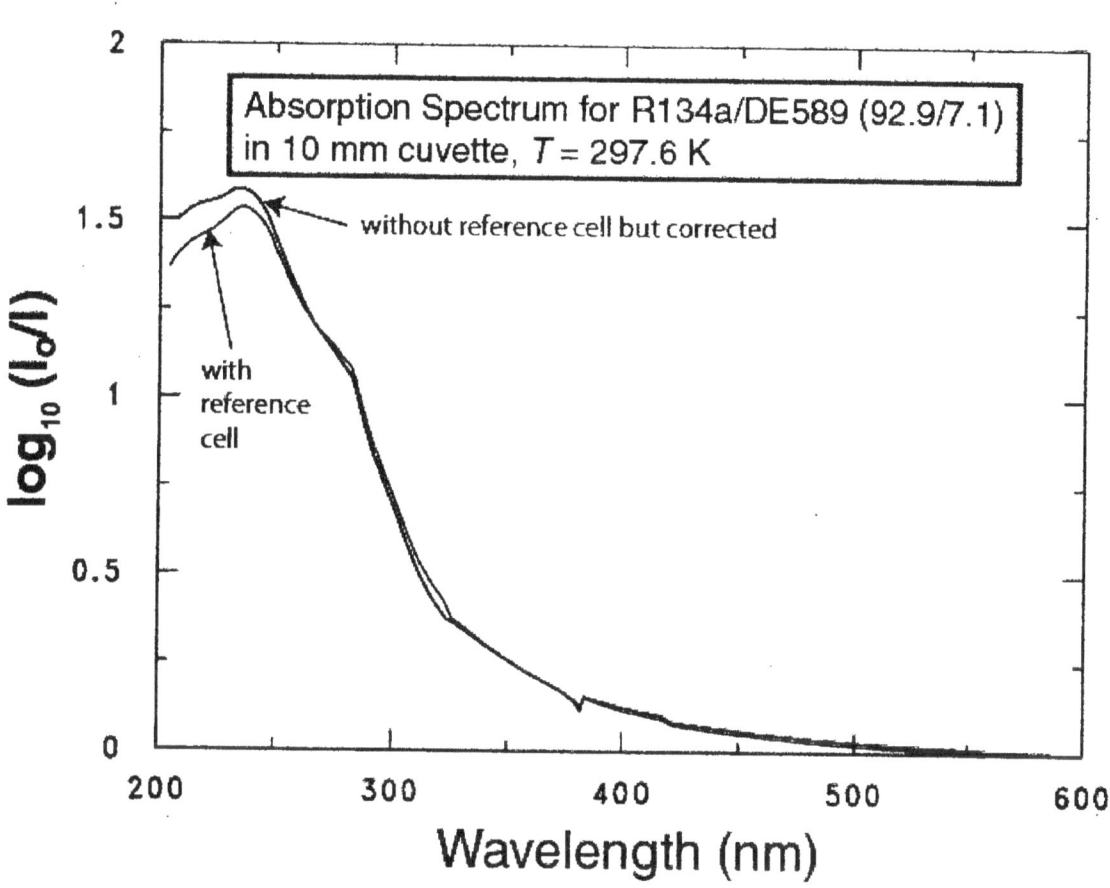

Fig. H.1 Absorption spectrum for R134a/DE589 (92.9/7.1) mixture

APPENDIX I

This appendix shows the pure R134a heat flux versus superheat for "surface aged" and prior to "surface aged" data. The data using the filled circle, square and triangle were deemed to be surface-aging data. The remaining data taken after August 23, 2001 were considered to be for an aged-surface. Only the aged-surface data were compared to the mixture data and uncertainties were given for only this data.

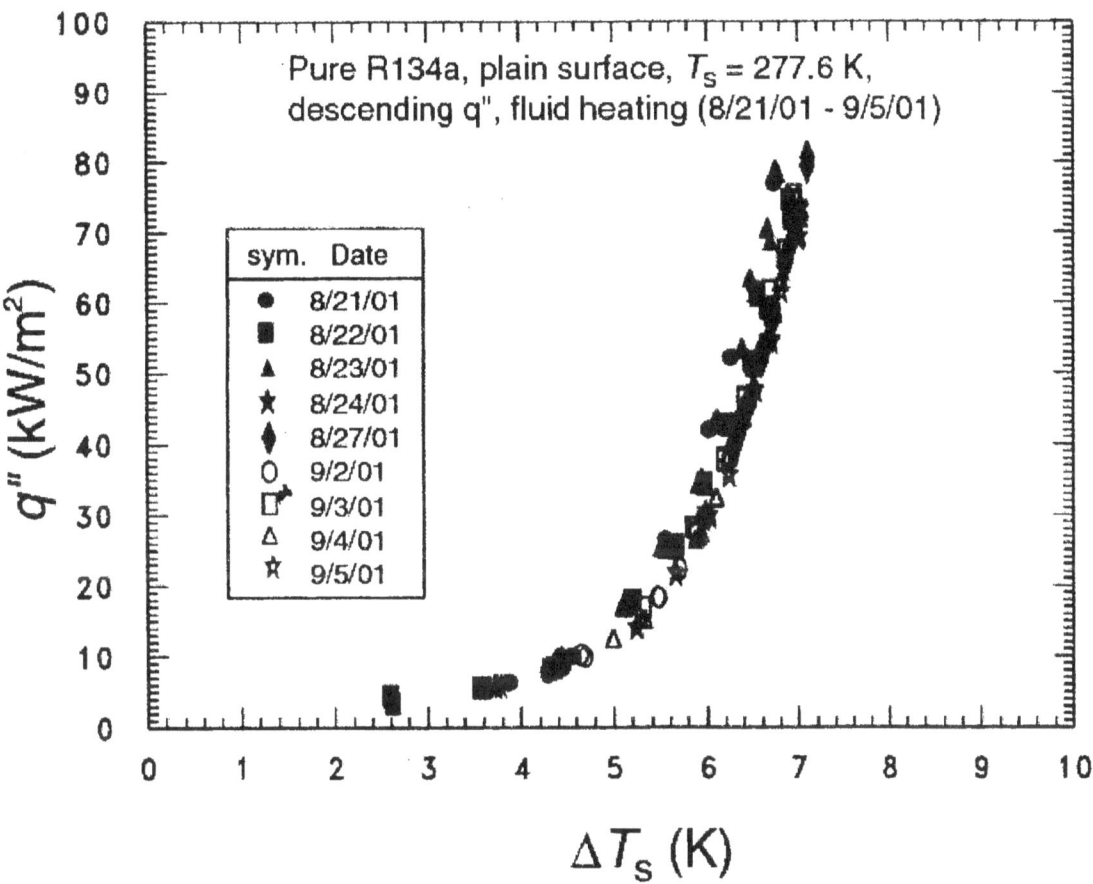

Figure I.1 Pure R134a pool boiling data with "prior to surface aging" data

APPENDIX J

This appendix derives the modification to the dimensionless lubricant excess surface density number presented in Kedzierski (2003) so that it accounts for both fluid pressure and variable ζ (concentration effects as it affects lubricant removal). These modifications were necessary so that the dimensionless Γ could be used to correlate both R123/York-C and R134a/DE589 measurements: mixtures of two different lubricants with refrigerants of significantly different reduced pressures ($P_r = P/P_c$) at the test temperature.

The expression for the bubble radius (r_b) as a function of the lubricant mass fraction (x_b) is given in Kedzierski (2003) as:

$$r_b = \frac{0.75 \zeta l_e \rho_L (1-x_b)}{x_b \rho_{rv}} = \frac{18.75 \text{Å} \rho_L (1-x_b)}{x_b \rho_{rv}} \tag{J.1}$$

Semeria (1962) and Nishikawa and Urakwa (1960) have shown that the departure bubble radius is proportional to the reduced pressure raised to the -0.5 and -0.6 powers, respectively. Consequently, one way to correct Eq. (J.1) for the effects of reduced pressure would be to multiply it by KP_r^n where K and n are both constants.

Setting the modified expression for the bubble radius to that given by the Clausius-Clapeyron equation begets:

$$r_b = \frac{0.75 K P_r^{-n} \zeta l_e \rho_L (1-x_b)}{x_b \rho_{rv}} = \frac{2 T_s \sigma}{h_{fg} \rho_v \Delta T_s} \tag{J.2}$$

Note that it is assumed that the Clausius-Clapeyron equation captures the effects of pressure via the variation in fluid properties and the absolute saturation temperature T_s.

Solving Eq. (J.2) for the lubricant excess surface density thickness (l_e) yields:

$$l_e = \frac{8 T_s \sigma x_b}{3 K \zeta P_r^{-n} l_e \rho_L (1-x_b) h_{fg} \Delta T_s} \tag{J.3}$$

The definition of the excess layer $l_e = \Gamma / (\rho_L - \rho_b x_b)$ can be used to write Eq. (J.3) in terms of the lubricant excess surface density (Γ):

$$\Gamma = \frac{8}{3 K \zeta} \frac{(\rho_L - \rho_b x_b) x_b T_s \sigma P_r^n}{(1-x_b) \rho_L h_{fg} \Delta T_s} \tag{J.4}$$

As reported in Kedzierski (2003), ζ is the fraction of l_e that is removed from the wall by the bubble in a disk of radius r_b and thickness ζl_e. In the previous study, variation of the lubricant thickness on the bubble with respect to the bulk mass fraction was not determined. Instead, it was assumed that the lubricant thickness on the bubble was constant. Presently, it is assumed that ζ is related to the bulk lubricant mass fraction as:

$$\zeta = \gamma x_b^{-p} \tag{J.5}$$

where both γ and p are constants.

Substitution of Eq. (J.5) into Eq. (J.4) gives:

$$\Gamma = \frac{8}{3K\gamma x_b^{-p}} \frac{(\rho_L - \rho_b x_b) x_b T_s \sigma P_r^n}{(1 - x_b) \rho_L h_{fg} \Delta T_s} \tag{J.6}$$

The excess surface density measurements (Γ) were linearly regressed against the parameters on the right-side of Eq. (J.6). The values of the exponents n and p that were obtained from the regression where 0.8 and 1, respectively. Grouping the leading constants into a single constant D and making use of the regression results for the exponents gives:

$$\Gamma = D \frac{(\rho_L - \rho_b x_b) x_b^{1.8} T_s \sigma P_r}{(1 - x_b) \rho_L h_{fg} \Delta T_s} \tag{J.7}$$

Regression of Γ against the parameters following D in Eq. (J.7) yielded $3.6 \times 10^9 \pm 0.28 \times 10^9$ for the value of D.

Equation (J.7) can be rearranged to obtain the new constant for the modified dimensionless Γ which accounts for the effects of reduced pressure and variable ζ:

$$\frac{(\rho_L - \rho_b x_b) x_b^{1.8} T_s \sigma P_r}{(1 - x_b) \rho_L h_{fg} \Delta T_s \Gamma} = 2.8 \times 10^{-10} \pm 0.2 \times 10^{-10} \tag{J.8}$$

As a check on the derived functional relationship between ζ and x_b, the thickness of the lubricant on the bubble (l_a) is averaged over the range of lubricant mass fractions tested and set equal to the constant for l_a that was assumed in the Kedzierski (2003) analysis:

$$\bar{l}_a = 25\text{Å} = \frac{1}{x_{b1} - x_{b2}} \int_{x_{b1}}^{x_{b2}} l_e \zeta \, dx_b = \frac{1}{x_{b1} - x_{b2}} \int_{x_{b1}}^{x_{b2}} l_e \gamma x_b^{-0.8} \, dx_b \tag{J.9}$$

The thickness of the excess layer on the wall is equal to a property group C multiplied by $x_b/(1-x_b)$. Considering that the integration is from $x_{b1} = 0.005$ to $x_{b2} = 0.02$, the analysis is done with $l_e = Cx_b$ resulting in the following expression for l_a:

$$l_a = 60.58 \, \text{Å} \, x_b^{0.2} \tag{J.10}$$

Equation (J.10) makes physical sense and justifies the constant value of 25 Å for l_a because the relatively small exponent on x_b does not cause a significant variation of l_a from $x_b = 0.005$ ($l_a = 21.0$Å) to $x_b = 0.02$ ($l_a = 27.7$Å).

www.ingramcontent.com/pod-product-compliance
Lightning Source LLC
Chambersburg PA
CBHW081738170526
45167CB00009B/3860